A Victorian Scientist
and Engineer

The young engineer: Fleeming Jenkin, *c.* 1860
(Jenkin mss)

A Victorian Scientist and Engineer

Fleeming Jenkin and the Birth of Electrical Engineering

Gillian Cookson and Colin A. Hempstead

Ashgate

Aldershot • Brookfield USA • Singapore • Sydney

Published by

Ashgate Publishing Ltd
Gower House, Croft Road,
Aldershot, Hampshire GU11 3HR
England

Ashgate Publishing Company
Old Post Road
Brookfield, Vermont 05036–9704
USA

Ashgate website: http://www.ashgate.com

ISBN 0 7546 0079 3

British Library Cataloguing-in-Publication Data
Cookson, Gillian
    A Victorian Scientist and Engineer: Fleeming Jenkin and the Birth of
    Electrical Engineering
    1. Jenkin, Fleeming, 1833–1885. 2. Electrical engineering—Great Britain—
    History—19th century. 3. Electrical engineers—Great Britain—Biography.
    I. Title. II. Hempstead, Colin
    621.3'092

US Library of Congress Cataloging-in-Publication Data
Cookson, Gillian
    A Victorian Scientist and Engineer: Fleeming Jenkin and the Birth of
    Electrical Engineering
        p.   cm.   Includes bibliographical references (p.)
    1. Jenkin, Fleeming, 1833–1885. 2. Scientists—Great Britain—Biography.
    3. Engineers—Great Britain—Biography. I. Hempstead, Colin. II. Title.
    Q143.J5H46   2000   99–41823
    509'.2–dc21   CIP

This volume is printed on acid-free paper

Printed and bound by Athenaeum Press, Ltd.,
Gateshead, Tyne & Wear.

Contents

Foreword

The Right Honourable the Lord Jenkin of Roding

On a recent visit to the Institution of Civil Engineers I found myself gazing up at the portrait there of William Thomson, Lord Kelvin. It is typical of its genre – dark, formal, dignified, imposing; the tall, gaunt, bearded figure gazes down on the lesser mortals around him. The painter has caught not only the distinction of one of the greatest of the Victorian engineers but also the formidable authority of the man.

Yet having read and reread this biography of Thomson's collaborator and business partner, my great-grandfather, Professor Fleeming Jenkin, I could not avoid seeing also something of the self-regarding pride which seems to have made it difficult for him to give the younger man full credit for the latter's undoubted achievements. Indeed, as Dr Cookson and Dr Hempstead recount, Thomson's obituary of his recently deceased partner, written for the Royal Society, reflected his view 'that any importance which attached to Jenkin flowed only from his relationship with Thomson, a relationship in which Jenkin was very much subordinate'. Compared with the tributes to be found in other appreciations of Jenkin's life of achievement (not least the Institution of Civil Engineer's own eloquent obituary) this unwillingness to recognise what other contemporary engineers gladly acknowledged may partly account for Jenkin being regarded for many years as a somewhat obscure, minor figure.

There is another, quite different, reason why it is not until recently that Jenkin's contributions to the science of electromagnetism have not been more widely recognised. The only previous record of his life before this new biography was the Memoir written soon after his death in 1885 by his friend, Robert Louis Stevenson. As Cookson and Hempstead make clear, Stevenson had no love for and little interest in electrical engineering. Indeed, some of the major discoveries of the era receive no mention whatever in the Memoir, let alone Jenkin's seminal contributions to those discoveries. Yet the Memoir is a lovely book, full of all sorts of reminiscences and events, capturing much of the extraordinary personality of his dear friend. My copy has been well thumbed by successive generations of the Jenkin family.

It was not until I was invited to contribute to a symposium on 'Clerk Maxwell and his Circle' in 1995, that I began to find out just why Fleeming

Jenkin should be seen as one of the leading engineers of his age. As I delved for the first time into the biographies of such eminent men as Sir William Thomson, later Lord Kelvin, P.G. Tait, Sir Alfred Ewing and, of course, James Clerk Maxwell himself, I found myself agreeing with Sir Alfred Ewing, who was perhaps Fleeming Jenkin's most distinguished pupil, that 'Fleeming Jenkin was not well served by his biographer'. This new biography by Cookson and Hempstead more than makes up for the shortcomings of the earlier work, and does so comprehensively and with considerable authority. The reader needs only to glance through the copious footnotes and references to appreciate that a deal of painstaking research lies behind this new work. Yet, dare I suggest, the reader who wishes to understand just why Fleeming Jenkin attracted such wide affection and admiration from his circle of friends and colleagues, and why his premature death from septicaemia caused such deep sadness, needs to read both books. This is not meant as any criticism of the authors' work; it is simply that a modern history, however carefully researched, cannot capture the emotions of those who knew the man intimately and admired the complex personality that lay behind his eventful career.

It is not the purpose of this brief Foreword to attempt a critique of the book. I am content to let readers make their own judgements. What I seek to do is to draw out a few threads from my great-grandfather's life that have a relevance to the modern world. Of course, technology has moved forward in ways which he and his contemporaries would find incredible: the transistor and silicon chip, wireless communications, opto-electronics, and advances into space and satellite communications – all these have created a world of which nineteenth-century engineers could scarcely have dreamed. If one throws in nuclear fission, particle physics and radio astronomy, it would be easy to imagine that there could be little or nothing in common between those early telegraph pioneers and today's generation of scientists and engineers.

But that would be to assume that we have nothing learn from history. As I read this book, I was struck again and again by how much there is that is relevant to us today. For instance, the relationship between industry and academia, a very contemporary issue, was an important theme in Fleeming Jenkin's life. Then, as now, in a fast moving technological environment commercial developments ran far ahead of the generality of university teaching. Industry was leading in the science as well as in its application, and the universities were hard pressed to keep up. A parallel today is the Internet and the World Wide Web, where most universities can do little more than try to keep abreast of what is happening in industry.

For many engineers who read this book, the section on the education and training of engineers will be of great interest. Fleeming Jenkin, coming from industry into academia, wanted his students both to have a firm grounding in the *theoretical* science of the subject, and to acquire the *practical skills* they would need in industry. Jenkin's philosophy of engineering education lives on today in many universities. The terms of his employment allowed him to continue with his consultancies, and some of his most notable achievements were made after he took up his appointment in Edinburgh. In particular, what some regard as Jenkin's chief claim to an honoured place in the history of science, his work for the British Association's Committee on Electrical Standards led to the adoption of standards of absolute electrical measurement – the ohm, the ampere, and other standards – which are in general use across the world.

Jenkin's own unconventional education, which resulted in fluency in three continental languages and brought him exposure to different cultures, traditions and educational experiences, also has lessons for us. It is far too difficult today for an engineering student to undertake the different stages of his or her education in more than one country. There may be 'Fleeming Jenkins' out there but the rigidities of syllabuses make it much harder to take part of one's training abroad than it was in his day. Is it too much to hope that the European Union may at least take us back to the flexibilities of a century-and-a-half ago?

For me, therefore, reading this book has been much more than dry history or a family annal. There are here all sorts of lessons and insights relevant to our state from which we may perhaps draw some wisdom to guide us in the future. The authors have done their work well, and very thoroughly, and readers will find much to amuse, inform and enlighten them.

I must end with two coincidences of purely personal interest. When my great-grandfather was chosen in 1868 to be the first holder of the new Chair of Engineering at the University of Edinburgh, among the six Edinburgh professors who formed a committee to recommend the appointment was the distinguished Professor of Materia Medica of the day, Professor Robert Christison (later Sir Robert). In 1952, in St Mary's Cathedral in Edinburgh, I married Sir Robert's great-granddaughter.

Second, when, also in 1868, the government of the day decided that the network of commercial telegraph cables within the jurisdiction of parliament, both overland and submarine, should be nationalised, it was my great-grandfather who was charged with the task of testing those cables and reporting on their condition to the government. One hundred and fourteen years later, in 1982, as Secretary of State for Industry, I introduced a Bill into the House of Commons to privatise British Telecom and so return the successors of those cables to the private sector. I hope that my enterprising ancestor approved!

Patrick Jenkin
May 1999

Acknowledgements

During the course of our research on Fleeming Jenkin and submarine telegraphy, we have received assistance from many archivists, librarians and museum staff in England, Scotland and the United States, for which we are extremely grateful. In particular we would like to thank Tim Procter, formerly of the Institution of Electrical Engineers archives, for taking an interest which went far beyond the call of duty; Sheila Mackenzie of the National Library of Scotland; Elaine Greig, curator of The Writers' Museum, Edinburgh; Carol Arrowsmith of the Institution of Civil Engineers archives; Keith Moore of the Institution of Mechanical Engineers; Lenore Symons and Anne Locker of the Institution of Electrical Engineers archives; Brian Cook, secretary of the Edinburgh Academical Club, and Alan Jarman, archivist to the Edinburgh Academy, who went to great lengths in answering questions about Jenkin's school career. Alison Morrison-Low, keeper of scientific instruments at the Royal Museum of Scotland, has been a mine of information about Jenkin's life in Edinburgh and about instrument making, as well as commenting on draft chapters and providing unforgettable Scottish hospitality. Susan Morris has been generous with her knowledge of Jenkin's work in economics; Norton Sloan shared his ideas on the relationship between William Thomson and Jenkin; Peter Filby enlightened us about Jenkin's teaching at University College, London; John Varley Jeffery has given us insights into the life and work of Cromwell Fleetwood Varley; Ben Marsden provided useful leads on source material. The Reverend Father David Hancock, parish priest of Stowting, Kent, devoted time on our behalf to finding details of Jenkin family burials. Jan Hewitt and Hugh Torrens have kindly offered ideas on aspects of Jenkin's life. We are also grateful for the advice and support of John Smedley and his colleagues at Ashgate Publishing Ltd.

Fleeming Jenkin's descendants, in particular Lieutenant-Commander James Jenkin, a great-grandson of Frewen Jenkin and custodian of the family papers, have dealt patiently with our enquiries and entrusted us with photographs for copying. We are grateful for the support and encouragement of Lieutenant-Commander Jenkin, and of the Right Honourable the Lord Jenkin of Roding.

The research on Fleeming Jenkin and writing of this book would have been impossible without the generous financial support of the Leverhulme Trust, and their help is gratefully acknowledged.

Finally, recognising that they too have had to live with Fleeming Jenkin, we would like to thank our families for their continuing patience and encouragement, and to mention in particular our debt to Jean Hempstead and Neil Cookson.

GC and CAH
May 1999

List of Figures

Figures 4.2, 4.3 and 4.7 are reproduced with permission from originals held in the Archives Department of the Institution of Electrical Engineers; pictures from originals in a collection of Jenkin manuscripts are printed with the permission of the Right Honourable Lord Jenkin of Roding.

Abbreviations in Footnotes

Jenkin mss   Jenkin family papers, in the possession of Lieutenant-Commander James Jenkin, R.N., Binsted, Hampshire.

Kelvin, Cambridge   Cambridge University Library, Kelvin collection, Add.7342.

Kelvin, Glasgow,   Glasgow University Library, Kelvin papers.

PRO   Public Record Office

RLS   R.L. Stevenson, 'Memoir of Fleeming Jenkin' in Colvin and Ewing, (eds), 1887, I, xi–cliv.

Chapter 1

Introduction

'I read my engineers' lives steadily, but find biographies depressing'[1]

In a life of only fifty-two years, FJ established his reputation as a pioneer in the new world of electrical engineering, in particular for his work on undersea telegraphs. Equally at ease in realms of theory and practice, from 1850 until his death in 1885 Jenkin engaged in every field of Victorian engineering. Never one to follow 'the beaten track' – his own words – from his youth he sought the chance to be 'a real designer able to carry out new and difficult work'.[2] He found his opportunity in intercontinental submarine telegraphy, the cutting edge technology of its day which progressed in tandem with the new science of electricity. Jenkin was both a scientist and an engineer, a prototype of the modern experimental research engineer. He was also a distinguished academic, professor of engineering in the University of Edinburgh, admired as an inspired and innovative teacher, and for his interest in the philosophical tenets underpinning his subject. His output included numerous scientific papers, thirty-five British patents, popular articles on science and technology, and influential basic textbooks. Not least, Jenkin played a crucial role in developing a world-wide standard of electrical measurement.[3]

The story of Jenkin's life provides a window into the world of Victorian engineering, and especially into the foundation of the new, scientifically-based branch of electrical engineering. More than most engineers, Jenkin understood theory as well as practice, and was concerned to develop electrical science alongside technology. Although slightly junior to the very young engineers responsible for the first attempts to lay an Atlantic cable in 1857 and 1858, Jenkin qualifies as a first generation electrical engineer, his main work falling in the middle years of Victoria's reign. Most electrical discoveries and technologies of the nineteenth century happened during his lifetime, and the scientific bases of many twentieth-century electrical technologies were in place when he died.

---

[1]   Fleeming Jenkin, quoted by RLS, p. cxlv.
[2]   University of St Andrews Library, Forbes correspondence (in), 1865/62, Jenkin to Thomson 5 July 1865.
[3]   Hempstead 1998, pp. 35–6.

The year 1799, when Volta's pile of zinc and copper plates separated by dampened paper pointed the way to producing relatively large amounts of low tension electricity, marks the beginning of electricity as a modern science. Following Volta, Cruikshank constructed a battery which could produce continuous and long-lasting currents. The Daniell cell, which appeared in 1836, generated a more constant electromotive force, required less maintenance, and was to provide the power for telegraphs until replaced by safer and more reliable sources in the second half of the nineteenth century.

The Voltaic pile was first used in 1800 to 'electrolyse' water into its constituent gases, and by 1807 Humphry Davy was using large Cruikshank type batteries to isolate potassium and sodium. Electrolysis had been brought into use in industry by the 1840s in electrolytic plating. Electrolytic copper refining, resulting in a far purer product than could be obtained by chemical processing, started as a commercial operation in 1869, to satisfy the demands of the submarine telegraph system for large quantities of high quality copper.

Oersted, working with Voltaic piles, discovered in 1820 that electric currents generated magnetic fields. From this physical discovery came a development in technology, the design and construction of electromagnets, small versions of which were later used in relays and receiving devices in electric telegraph systems.

Electric telegraphs had first been suggested at the end of the 18th century. Sir Francis Ronalds built an experimental telegraph in his garden in about 1816 using static electricity. In the 1830s Charles Wheatstone was able to develop a commercially viable telegraph using Oersted's discovery of the magnetic effect of an electric current.[4] Wheatstone's 1837 telegraph patent with William Fothergill Cooke is generally viewed as a milestone, although similar systems originated in other parts of Western Europe and the United States during the 1830s. All practical devices minimised the number of conductors, eventually by using earth returns with only a single wire. Most had to employ some form of code, although there were strenuous efforts to design machines which would translate signals into letters for the operators. On early telegraphs messages were sent by hand and read letter by letter, a laborious and unreliable process. The first printing telegraph was designed by Alexander Bain in 1840, and was followed by many other versions.

By 1856, just before Fleeming Jenkin became an electrical engineer, the first submarine cables across the English Channel, from Dover to Calais and between Orford Ness and the Hague, had brought the major population centres of western Europe into direct contact with the cities of the United Kingdom. To cope with rapid increases in telegraph traffic, duplex telegraphy, or

---

[4]   See Brian Bowers, *Sir Charles Wheatstone, FRS, 1802–1875* (London: Science Museum, 1975), especially ch. 9.

simultaneous transmission in opposite directions, was developed. A first method, using synchronous revolving switches or commutators and suggested by Moses Farmer of Boston in 1852, proved technically problematic. Following experiments with an electrical duplex system on a line between Prague and Vienna in 1853, a commercially viable duplex telegraph emerged the following year.

Vital to the construction and working of long-distance telegraphs was the development of testing equipment, which in its turn was a product of scientific research. In detecting breaks and faults, engineers were aided by 'remote' methods of measurement which relied on a knowledge of the behaviour of electric currents in circuits and resistances, and demanded methods of comparing electromotive forces and detecting currents. The electromagnetic discoveries of Oersted allowed the construction of galvanometers, and the Wheatstone bridge was a convenient comparator.[5]

When Jenkin joined Newall in 1857, elaborate senders and receivers, relays and duplexers were in common use on many thousands of miles of land telegraphs, and short submarine telegraph lines were working more or less reliably. The technology of terrestrial telegraphy was then well understood – and did not attract Jenkin's interest – but submarine telegraphy, with the failure of the 1857 attempt to cross the Atlantic, had suffered the first of several major setbacks. Some of the problems were managerial, or financial, or mechanical, or navigational, but it had also become clear that knowledge of the science of telegraphy was far from adequate. Without this understanding, long submarine lines would not be installed and operated successfully. Many years of electrical experimentation had not brought reliable quantification, nor were there theories to help telegraph engineers. Jenkin was to be one of the scientist-engineers who would become founders of the new science of electricity.

Yet the story of Jenkin's life also shows that he was far more than an engineer and scientist of great range and originality. He had many other interests, and he engaged in them with extraordinary depth and effect. In the spheres of economics and social issues, on the construction and staging of drama, on Darwin's evolutionary theory, on philosophy, Jenkin's contributions were at the very least valuable, and were sometimes remarkable. His achievements were born of an 'unresting, insurgent intellectual activity' tempered by deep humanity.[6] His friend and former student Alfred Ewing remembered Jenkin as an eminent engineer and inspiring teacher, and much

---

[5]   This device allowed four resistances A, B, C, D to be connected in a diamond pattern. Between one pair of opposite vertices a galvanometer was connected, across the other pair there was a battery. If $A/B = D/C$ then there was no current flowing through the galvanometer, hence if three resistances were given then under these, balance conditions the value of an unknown could be measured.

[6]   Stevenson, obituary in *The Academy*, 20 June 1885.

more besides: 'His interest in art, in literature, in personalities, in all that makes up life, was unbounded; his judgement was penetrating and sympathetic.' [7]

Jenkin's collaborators and acquaintances ranged from some of the most distinguished scientists and engineers of the day, to several of the best known novelists and political economists. The engineers with whom he associated included William Thomson, James Clerk Maxwell, William Fairbairn and Josiah Latimer Clark. The authors Elizabeth Gaskell and Robert Louis Stevenson were among his close friends. To the latter, Jenkin was a much loved mentor, his friendship doing much 'to temper with sweetness and sanity Stevenson's early years of struggle and revolt'.[8]

Yet in spite of his influence as an early electrical engineer and his other intellectual achievements, despite the celebrity of his associates and the way that submarine telegraphs seized the Victorian popular imagination, Jenkin himself has remained an obscure figure. It is not unusual that a person's fame bears little relationship to the true quality of their achievement, and Jenkin deserves to be better known. The kind of work in which he interested himself partly explains why he has not been more fully recognised and appreciated. His character also told against him, for he managed to alienate people who did not know him well by his direct and challenging manner, while being naturally modest and so concerned to give others their due that he neglected to promote his own reputation in the way of some of his contemporaries.[9] He died suddenly in middle age, and the memoir of him which Stevenson wrote established an image which did not do full justice to Jenkin himself, nor to the scale of his achievement.

Submarine telegraphy – connecting Britain with the continent in 1851, and Europe with America during the mid 1860s – was of enormous interest to the general public during Jenkin's early career in electricity. Jenkin, though, was still a junior engineer working in the background when William Thomson and Charles Bright, only slightly older men than he, achieved fame on the highly publicised, though unsuccessful, attempts to span the Atlantic in the late 1850s. Jenkin's most important electrical work, on standards and measurements, was too theoretical to have had much general appeal. He was also somewhat overshadowed by Thomson, his closest professional collaborator in a relationship which lasted for more than twenty five years. Jenkin's open admiration for Thomson, and self-effacement in his relations with the famous man, tended to diminish Jenkin's own stature in the eyes of Thomson and the wider world.

---

[7] Ewing 1933, p. 248.
[8] Ewing 1933, p. 249. For Stevenson's relationship with the Jenkins, see Ewing, ch. X.
[9] For a more complete discussion of this, see Cookson 1998.

Jenkin was not a calculating man. Although he made firm and loyal friends, he never deliberately sought out those who might advance his career or his standing. The pursuit of truth was too important. If accolades or recognition came, then he welcomed them, but he did not court approval. His school friend and later colleague Peter Guthrie Tait emphasised Jenkin's sincerity, and his bluntness: 'The late Prof. Jenkin was essentially a frank, straightforward, hard-working, clear-headed practical scientific man – and it is in this capacity that he will be held in honourable remembrance by the scientific world'.[10] Above all, Jenkin had 'sterling honesty'. If some of his writings appeared forceful and opinionated, the reason was not, according to Tait,

> the self-laudation of a flippant 'paper scientist'; it is the deliberate statement of a clear-headed man who took nothing for granted, and who never wrote on anything till he felt convinced that he understood it.[11]

Tait here touches on those qualities of impatience and directness which did not please everyone and which positively militated against a career as an heroic Victorian engineer. Yet while Jenkin never set out to charm, those who knew him well would have thought harsh, perhaps to the point of inaccuracy, this summary of his character in the *Dictionary of National Biography*: 'Very plain-featured, rather short in stature, always youthful and energetic in manner, Jenkin did not prepossess strangers, and his flow of words and disputation never made him very popular'.[12] Friends produced a more balanced summary of his character. 'In talk he was active, combative, pounced upon his interlocutors, and equally enjoyed a victory or a defeat. He had both wit and humour; had a great tolerance for men, little for opinions; gave much offence, never took any', said Stevenson. 'He was of the most radiant honesty and essentially simple; hating the shadow of a lie in himself, loving the truth, however hard, from others'. But to those who did not know him well, added Stevenson, 'to people of weak nerves or of a vulnerable vanity, he was at times a trial'.[13] Jenkin, while challenging, was also entertaining company, 'fonder of pure nonsense than [was Anne, his wife]; and very ingenious, witty and amusing'.[14] Jenkin's long-time friend Sidney Colvin, man of letters and Keeper of Prints and Drawings at the British Museum, said of him:

> Keenness shone visibly from his looks, which were not handsome but in the highest degree animated, sparkling and engaging, the very warts on

---

[10] Tait 1888, p. 434.
[11] Tait 1888, p. 435.
[12] Volume XXIX (1892), p. 296.
[13] Stevenson, obituary in *The Academy*, 20 June 1885.
[14] Jenkin mss, Mrs Roscoe's recollections of Anne Jenkin.

his countenance seeming to heighten the vivacity of its expression. The
amount of his vital energy was extraordinary ...[15]

So while Jenkin's friends, as one would expect, admired him and were
sympathetic to his foibles, some of his academic colleagues had much more
difficulty in accommodating his prickly nature and the directness in his
character. Although accounts diverge sharply when considering the degree to
which Jenkin was lovable, there is agreement that he made a serious and
significant contribution across a range of activities, that he was able and honest,
and above all that his motivation came from a desire to establish truth and
appreciate beauty. But his lack of concern for gaining approval, seeking
powerful allies or insisting upon due recognition as an innovator, worked
against his advancement into the communion of 'engineering saints', the group
canonised by Smiles and others.

By his sudden and untimely death in 1885, Jenkin was also denied a
position as a grand old man of engineering, with the honours and fame which
might have accompanied that role. Though many who lived far longer have
achieved a good deal less, there is a sense with Jenkin that much was still
unfinished, that greater work was to come. His early death also meant that his
attainments may not have been clearly recognised, for some of his colleagues
were subsequently less than generous in giving Jenkin the credit he deserved.
In particular Sir William Thomson, who had relied heavily upon Jenkin to
manage the business aspects of their patent marketing partnership, appears to
have been more concerned to promote his own image than to acknowledge
Jenkin's contribution in his public accounts of Jenkin's career. History is
written by the victors, and that includes survivors, those who have outlived
their colleagues.

Yet there was nothing ungracious about Robert Louis Stevenson's
memoir of Jenkin, and that too has been the means of undermining Jenkin's
credibility as a major figure in science and engineering. The memoir is the only
full-length biography of Jenkin to date, and the only one Stevenson ever
completed. It is a curious, flawed work which may ultimately have done a
disservice to the friend its author sought to honour. In trying to persuade his
American publisher to take the memoir, Stevenson presented it as a fusion of
the kind of adventure story in which he excelled, with an account of the life of
a great engineer in the manner of Samuel Smiles: 'An odd man, an odd family
history, no lack of adventure or human interest; and the public likes engineers,
does it not?' but he also conceded that Jenkin was a 'not very exciting
novelty'.[16] Stevenson had for once missed the mark and failed to interest a

---

[15]   Colvin 1921, p. 157.
[16]   Booth and Mehew (eds) 1995, V, p. 357, Stevenson to Messrs. Charles Scribner's
Sons, 12 February 1887.

general readership while simultaneously antagonising Jenkin's more serious scientific colleagues. He was too close to his subject, it was too soon after the shock of his death, and in truth Stevenson, though well-intentioned, was not cut out for biography.[17] Nor was Jenkin, with his complex character and wealth of interests which made him difficult to pigeonhole, an obvious candidate to popularise as a 'Great Engineer'.

Still the Stevenson memoir does have value, for it has itself become part of Jenkin's story. Above all it is testimony to the author's love for his subject, confirming that a significant strand of Jenkin's achievement was the support, encouragement and friendship which was to prove instrumental in helping a sickly and troubled young Stevenson to become a great author. Stevenson had not intended to produce any balanced account of Jenkin's work as an engineer. He wanted to present the essence of Jenkin's character and life in a book which would be a lasting memorial. When he wrote, he was still reeling from the loss of Jenkin: 'Two years have passed since Fleeming was laid to rest ... and the thought and the look of our friend still haunts us'.[18]

The plan to write the memoir was agreed between Stevenson and Anne Jenkin within two weeks of Jenkin's death. Despite Anne's help with research, information and other practical matters, it was two years in the writing. In 1887 it was published as part of Jenkin's collected papers, edited by Sidney Colvin and Alfred Ewing.[19] The long delay was mainly caused by Stevenson's ailing health, and friends including Henry James had actively discouraged the project: 'I am a good deal distressed at hearing you are to collaborate with Mrs F. Jenkin. I doubt of your being in a state to tread a measure with a biographical widow; pray let the music be very slow'.[20] Stevenson was beginning to taste success as a novelist – *Treasure Island*, published in 1882, promised to be 'as rich as Cuba'[21] – but constant financial worries continued alongside his bad health.[22] He broke off from the biography during the autumn of 1885 in order to rush out, in ten weeks, *The Strange Case of Dr Jekyll and Mr Hyde*. Not until 1888 were his finances transformed, and he was able to describe himself to Anne Jenkin as 'Bloated with Lucre'.[23] From the circumstances and from

[17] Nor was he successful as a playwright: see Chapter 7, below.
[18] RLS, p. cliv.
[19] The memoir is referred to in footnotes as RLS. It can also be found in various collections of Stevenson's works.
[20] Yale, Beinecke, 4926, James to Stevenson, 8 December 1885.
[21] Booth and Mehew (eds) 1995, V, p. 198, Stevenson to Anne Jenkin, *c*.6 February 1886.
[22] He returned a much-needed publisher's advance of £100 during the writing of the memoir, as he feared that in the event of his death his family would have been unable to repay the money: Booth and Mehew (eds) 1995, V, p. 357, Stevenson to Charles Scribner's Sons, 12 February 1997.
[23] Booth and Mehew (eds) 1995, VI, p. 118, (?)late February 1888.

Stevenson's own comments it is clear that the memoir was always a labour of love: 'We were put here to do what service we can, for honour and not for hire'.[24] He was trying honestly to make sense of his friend's character and provide a record to honour him. 'And give me time, the picture, the face, will rise at last out of these botchings, or I am the more deceived; I do seem to see him clear and whole'.[25] As the work neared completion, he told Anne Jenkin: 'I really do think it admirably good. It has so much evoked Fleeming for myself that I found my conscience stirred just as it used to be after a serious talk with him; surely that means it is good?'[26]

The service which Stevenson did for Jenkin was to provide detail of a kind with which engineering biographers do not usually concern themselves. Thanks to Stevenson, we have a colourful account of Jenkin's life outside science and engineering, of his character, his motivations and his loves. But Stevenson inadvertently failed Jenkin in skating over the significance of his subject's achievements in science and engineering. The novelist, himself intended for a career in the family business of lighthouse construction and a former engineering student of Jenkin at Edinburgh, was competent to say more about Jenkin's professional life. Such was Stevenson's distaste for engineering, though, that his treatment of Jenkin's technical work was superficial, with enduring effects upon how seriously posterity has viewed Jenkin.

At the time of its publication and since, Stevenson's memoir has also been subjected to sharp criticism as a record of Jenkin's character and life outside engineering. Frank Swinnerton, an early biographer of Stevenson, thought Jenkin 'poorly drawn, so that he might be anybody'.[27] Tait criticised Stevenson for using a 'merciless scalpel', an editorial heavy hand which blurred Jenkin's true achievements.[28] Brander Matthews, who had also known Jenkin, believed that

> Jenkin was not altogether fortunate in his biographer, in spite of
> Stevenson's loyalty to his dead friend and in spite of his manifest desire
> to show this friend for what he was. Perhaps a narrative by a less expert
> pen would have presented a larger figure, or at least a figure less
> obscured by the personality of the biographer himself.[29]

---

[24] Booth and Mehew (eds) 1995, V, p. 172, Stevenson to Edmund Gosse, 2 January 1886.

[25] Booth and Mehew (eds) 1995, V, p. 342, Stevenson to Anne Jenkin, c.23 December 1886.

[26] Booth and Mehew (eds) 1995, V, p. 383, (?)7 April 1887.

[27] McLynn 1993, p. 249.

[28] Tait 1888, p. 434.

[29] Matthews (ed.) 1958, p. 72.

A more recent assessment of the memoir concludes that although 'charming' it tends to present Jenkin 'more as caricature than character'.[30] Certainly Stevenson's account of the Jenkin family history, in particular his description of Fleeming Jenkin's parents, was written from second-hand sources as an epic and exotic story which does not quite ring true. The secondary characters emerge as two-dimensional, or as eccentrics. As for his treatment of Jenkin himself, the accuracy of the portrayal is more open to question. Colvin, who knew Jenkin intimately, was flowery in his praise to Stevenson: 'Reading the memoir through continuously, I doubt if there is in literature a much more engaging, sound and salutary piece of human narrative ... 'tis a beautiful and loveable piece of work.'[31] The more restrained comments of Jenkin's middle son, Frewen, suggest that Stevenson had indeed captured something essential of his subject:

> I do not know how I can thank you for it. I think it is beautiful. It makes me feel more than ever what a friend I have lost, but it will help me in future to do what he would have liked. Some things you have understood and sympathised with more than I had thought possible for anyone but myself who am his son and inherit a little of his love of theories ... the whole book is so cheerful and lovely that I almost feel as if my father had come and told me the story of his life himself.[32]

The praise by those who knew Jenkin best suggest that the memoir can be used at least as a fair indication of his character. He was in reality a larger than life figure.

Stevenson's memorialising of Jenkin had another unfortunate effect. Many Victorian scientists and engineers of similar stature had biographies written by a close colleague or former student who would focus upon their technical achievements. In Jenkin's case, the task would probably have fallen to Alfred Ewing, a man well placed to describe experimental work and other activities during Jenkin's Edinburgh years.[33] Such a biography would have consolidated the scientific and engineering aspects of Jenkin's story, for there are no surviving notes of his experiments, nor sizeable collections of letters or papers. But Stevenson's intervention prevented any such scientific life appearing, and the opportunity to have Ewing's informed record of Jenkin's technical work after 1870 was missed. Since then, Stevenson's towering reputation as a writer may have deterred other potential biographers. Tait argued at the time that it was a great mistake to have had an artist as

---

[30] Morris 1994, p. 320.
[31] Yale, Beinecke, 4390, Colvin to Stevenson, 13 September [1887].
[32] Yale, Beinecke, 4982, C.F. Jenkin to Stevenson, 9 August 1887.
[33] Jenkin's own records of experiments have not survived.

biographer, although the kind of information he himself sought about Jenkin
was not necessarily of the widest appeal:

> Scientific men would have been glad to learn many things not mentioned
> here – e.g. the secret of his singularly methodical method of correspon-
> dence ... but his biographer is a true artist, for whom business, method,
> and even science itself have no attractions ...[34]

There is evidence that the editors of Jenkin's papers recognised that the
memoir would fall short in these respects, for they had already commissioned
appendices by Thomson on Jenkin's contribution to electrical science and
engineering, and by Lieutenant-Colonel Alexander Fergusson on Jenkin's work
on sanitary reform. These addenda briefly fill out aspects of Jenkin's life which
Stevenson preferred to avoid.[35]
     So while the memoir is dated in style, is open to question in many ways,
and has even adversely affected the reputation of its subject, it is also too
valuable a source of information to be ignored. It contains oral and
documentary evidence which has not otherwise survived, much of it in the form
of transcribed letters and journal entries by Jenkin himself. Stevenson may
have inflated aspects of Jenkin's character, but in general his account is rather
confirmed by others of Jenkin's friends in obituaries, reviews and private
letters. Stevenson's more personal feelings about Jenkin are also available to
us, in some of his own letters which are more heartfelt, not so calculated to
amuse, as the published work.[36] The relationship of the famous author and the
engineer, their love for each other and importance in each other's lives, is an
integral part of Jenkin's history.
     In Stevenson's memoir we have a unique record of the life of a
nineteenth-century scientist and engineer, albeit a very unusual specimen of
scientist with a remarkably catholic range of interests. For the purposes of this
present biography, the memoir has been treated in the main as a primary source.
Cross-checking against other available information shows that Stevenson was
overblown or vague – carelessly untroubled, for example, by dates or places –
rather than entirely inaccurate.
     In any case it is unfair to be too harsh on Stevenson for concentrating on
certain elements of his subject's life. Dealing with Jenkin's range of activities,
and in particular trying to assess the scale of his achievement in these different
fields, are challenges for any biographer. The hindsight of a century gives a

---

[34]   Tait 1888, p. 434.
[35]   It is clear from Stevenson's letters in 1885 and 1886 that it was always intended to
include these appendices; they were not grafted on later in response to a perceived shortfall in
the memoir.
[36]   See for example Booth and Mehew (eds) 1995, V, p. 172, Stevenson to Gosse, 2
January 1886.

clearer perspective on some parts of this. It becomes easier to identify the complementary aspects of Jenkin's wide range of work, to see consistency in apparently disparate activities, without constantly resorting to Jenkin's character to explain things, as Stevenson did.

Yet Jenkin's character shines through, and in particular his views of biography cast a long shadow over any attempt to deal with his own life. The reason he found biographies depressing was that:

> Misfortunes and trials can be graphically described, but happiness and the causes of happiness either cannot be or are not. A grand new branch of literature opens to my view: a drama in which people begin in a poor way and end, after getting gradually happier, in an ecstasy of enjoyment ... Smiles has not grasped my grand idea, and only shows a bitter struggle followed by a little respite before death.[37]

Jenkin's 'rainbow vision', his hope and optimism, deeply influenced Stevenson's approach to writing the life. Frewen Jenkin's view of the memoir as so 'cheerful and lovely' that it conjured up his father suggests that in this the author succeeded.[38] But maybe optimism had been emphasised at the expense of Jenkin's discernment and critical faculties, and it was this, the understatement of Jenkin's widely admired intellectual powers, which so offended some professional colleagues.

Before Jenkin can be recognised as a significant Victorian engineer and scientist, the image of superficiality which has surrounded him, based on Stevenson's description of his character and on the extraordinary compass of his interests, needs to be dispelled. The modern view is generally one of scepticism at the idea that a polymath could produce work of a consistently high, and sometimes extraordinary, quality over such a range of disciplines. The case of Jenkin shows that polymathy had not, as is sometimes claimed, expired in the eighteenth century. It was still acceptable in Jenkin's time for someone of his broad education to range widely around branches of science and engineering, so that his ideas about evolution and bridge construction, to give two examples, were received with respect. Then again, his contributions to economics, which were significant, were received publicly rather as unwelcome amateur efforts, yet were influential behind the scenes and found their way, unattributed, into the writings of leading economists of the time. Demarcation lines were appearing around new subject areas, and while some fields of expertise – like science and engineering – extended across a vast range, it was becoming less acceptable to stray beyond one's own specialisation. Jenkin did not allow such prejudices to constrain him, for

---

[37] RLS, p. cxlv.
[38] RLS, p. cliii.

narrowness was a characteristic he loathed. But he did show contradictions, a late polymath in the most modern branch of engineering, in an increasingly professional society, on a cusp between old and new. Jenkin bridged the classical world and the brave new world of electricity, and was interested in much between the two. 'There was no discussion in which he would not join, and no subject in which he did not take an interest; and such were his natural keenness of apprehension, and integrity and acuteness of judgement, that there seemed almost none on which he was not able to throw light,' said Colvin.[39] 'The variety and genuineness of Jenkin's intellectual interests proceeded in truth from the keenness and healthiness of his interest in life itself.'[40]

The story of Jenkin is of a life lived to the full. It illuminates many aspects of Victorian intellectual society, and of the organisation of science and engineering in that era. The central purpose of this biography is to show Jenkin's achievements in engineering and in other fields, and to judge his significance in these diverse activities. The content and style are intended to be accessible to a general reader. Those seeking more detailed scientific and technological information about the work of Jenkin and other early telegraph engineers should note that a complementary volume by the same authors, with a more technical focus, is planned to follow this.

---

[39] Jenkin's character is described at length in Colvin 1921, pp. 154–61; Colvin's account is partly reproduced in Lucas 1928, pp. 197–8.
[40] Colvin 1921, p. 157.

Chapter 2

A Young Man of Remarkable Ability[1]

The family of Fleeming Jenkin had deep roots in Kent. Generations of his ancestors had been well-to-do landowners in the county, and they settled in the manor of Stowting, between Ashford and Folkestone, from 1633.[2] Yet by the time Jenkin's father, Charles, was born at Stowting Court in 1801 the family had fallen upon hard times, said to have been the result of extravagance and impracticality. Their hope lay in a £60,000 legacy from a relative, Anne Frewen, wife of Admiral Buckner, referred to by Stevenson as 'the golden aunt'. When she died in the 1820s, leaving her fortune elsewhere, the family was almost ruined.[3] These circumstances influenced the course of Fleeming Jenkin's life, for although he was taken to Stowting at the beginning and end of his life, to be baptised and buried there, he was never able to live in the manor which his family considered their home, and was obliged to make his own way in the world. The temporary nature of Charles Jenkin's naval postings, and his enduring financial difficulties, brought an itinerant existence which profoundly affected his only child's upbringing. As a result, Fleeming Jenkin's education was broader and more cosmopolitan than that of most middle class boys of his time.

Even before the family's financial calamity, Charles Jenkin, as the second son, had not expected to manage the Kent estates. Admiral Buckner used his influence to help Charles into the Royal Naval College at Portsmouth at the age of thirteen. From 1817, for three years, Charles helped guard Napoleon in his exile on St Helena, a dreary duty which almost broke his health. Yet it was, in retrospect, the most noteworthy period of Charles Jenkin's forty-two years in the Royal Navy. The pinnacle of his naval career passed long before his son's birth. Serving in peacetime immediately after a long war, he found few chances

---

[1] This description of Jenkin was attributed by Thomson to Lewis Gordon in 1859 while Jenkin was working for Newall: Sir William Thomson, 'Note on the contribution of Fleeming Jenkin to electrical and engineering science', in Colvin and Ewing (eds) 1887, I, p. clv.

[2] Stevenson describes the family background and financial difficulties in colourful detail: RLS, ch. 1. He had access to Jenkin's own research into the family's history, carried out in about 1880: Jenkin mss.

[3] Stevenson gives conflicting dates – 1823 and 1825 – for the aunt's demise, but the essence of the story seems to be well-founded: RLS, pp. xv, xxv.

to distinguish himself. He was commissioned Lieutenant in 1829, but after promotion to Commander in 1846 was never again able to obtain employment.[4]

From 1823 until 1844, Charles Jenkin, Figure 2.1, was stationed chiefly in the West Indies, where in about 1828, while serving as mate on the *Barham*, he was introduced to the family of a fellow midshipman in Kingston, Jamaica. The man's sister, Henrietta Camilla Jackson, Figure 2.2, (born in 1808), daughter of the Honourable Robert Jackson, a planter and magistrate of Mahogany Vale, Jamaica, became his wife in 1831.[5] The Jacksons, along with

Figure 2.1: Henrietta Camilla
Jenkin, 1852 (Jenkin mss)

Figure 2.2: Captain Charles Jenkin,
c.1860 (Jenkin mss)

their mother's family, the Campbells of Auchenbreck, were of mainly Scottish extraction; like the Jenkins, they were in reduced circumstances. Otherwise the couple appear to have had little in common. Henrietta was high-spirited and artistic, with a particular talent for music and languages; Charles, by far the handsomer of the two, Stevenson portrays as a simple character who was loyal and devoted to his wife, but always in her shadow. Indeed, Stevenson's portrait of the captain is a caricature: though a gentleman, 'his mind was very largely blank' and he was used by Henrietta 'with a certain contempt'.[6] Following his father's death at about the time of the marriage, Charles bought with his wife's money a two-thirds share of Stowting, but the estate was so heavily encumbered that they derived no benefit from it until their old age.

---

    [4]   Jenkin mss, promotion order 23 December 1845. Charles Jenkin's full service record is given in PRO, ADM 196/1, p. 164. See also ADM 196/36, p. 693.
    [5]   *The Scotsman*, 9 and 10 February 1885.
    [6]   RLS, p. xxv.

Henry Charles Fleeming Jenkin was born on the twenty-fifth of March 1833 in a government building near Dungeness, where his father was serving in the Coast Guard.[7] He was always known as Fleeming (pronounced Flemming) Jenkin, named after his father's superior officer, the Honourable Charles Elphinstone Fleeming, Vice Admiral at Sheerness.[8] The family, which included his aunt Anna Jenkin and at times his grandmother Jackson, soon moved to Fort Sutherland, near Hythe, from where the baby was taken to Stowting for baptism. The following year, they removed to Vale Cottage, a mile from Charles Jenkin's Coast Guard Station at Sconce Point on the Isle of Wight. At the age of four, in 1837, Jenkin was placed for a year in the care of his grandmother while his mother and aunt accompanied Charles on a posting to the West Indies. Lieutenant Jenkin was in command of HMS *Romney*, a hulk moored in Havanna harbour. African slaves liberated from Spanish traders were temporarily accommodated on the *Romney*, and Charles Jenkin walked a careful path to carry out his orders without offending the local Spanish authorities.[9]

After their return from Cuba, the family lived at twenty-eight Pembroke Square, Kensington. Fleeming Jenkin was introduced to fishing and shooting, and learned to ride an old grey pony, on a holiday to the Scottish borders in 1840. The following year the house in Kensington was given up altogether in favour of a rented sporting estate at Hunt Hill House, near Jedbergh. It was here that Jenkin began his formal education as a day boy at Dr Burnet's Academy, Jedbergh, also called the Nest Academy. The school offered a range of subjects, including arithmetic, practical mathematics and book-keeping, geography, French and classical languages, along with some less conventional activities – drawing, music and dancing.[10] It is not known which of these subjects was studied by Jenkin, who at eight was one of the younger boys at the academy. But he was at most a year in Jedbergh, for when his father returned to sea in the West Indies, the rest of the family moved on again, to 'a fine old mansion', Barjarg Tower in Dumfriesshire.[11] Here his mother resumed full

---

[7]   Many of the details of Fleeming Jenkin's early life are taken from an account by his father in the Jenkin family papers. Stevenson used this selectively in his memoir of Jenkin. Charles Jenkin served as Lieutenant in the coastguard from November 1832 until May 1837: PRO, ADM 196/1, p. 164

[8]   Fleeming ended his career as an Admiral of the Blue, to which rank he was promoted in 1837. He disappears from Navy Lists in 1840.

[9]   There was at least one occasion during his command in Cuba when a diplomatic incident was narrowly avoided. Charles Jenkin chose to surrender an escaped Cuban slave to the authorities in Havana rather than risk the expulsion of the *Romney* and the collapse of the Mixed Slave Commission which aimed to suppress the slave trade. See Admiral Erskine's letter to the *Times*, 13 March 1876.

[10]   Garrett O'Brien, *The Nest Academy, Jedbergh* (Jedbergh: Crochet Factory, 1990).

[11]   Jenkin mss, father's notes. Charles Jenkin served on the *Avon* from September 1841 until the following April.

control of Jenkin's education, while allowing him to indulge his love of country sports and keep birds, ponies and other animals.

In 1843, the mother and aunt moved to Edinburgh for the express purpose of entering the ten year-old Fleeming Jenkin in the Edinburgh Academy, Figure 1.3. Though many boys from outside Edinburgh boarded with masters, Jenkin's family chose to take a furnished house in Northumberland Street. Jenkin was placed in Dr Cumming's class, a group of seventy-four pupils then in the third class, most of whom were two years older than he[12] – the choice of class was not dictated by age, but decided by parents.[13] The academy, founded in 1824, emphasised a classical education, yet there were modern elements in its curriculum. Not least of these was English as a compulsory subject, a radical innovation reflecting the influence of Sir Walter Scott as a founder of the school. While Latin grammar and literature occupied a central part of their

Figure 2.3: Edinburgh Academy (Y.Y., *Robert Louis Stevenson: A Bookman Extra Number,* London, Hodder and Stoughton, 1913)

learning, boys also studied Geography, Writing, and English, which included Ancient History and Scripture Biography.[14] From the 1830s there had been a growing recognition of the importance of mathematics, as a concession to a more practical type of education, and in the year that Jenkin joined the school, French, a language which he had already studied, became an optional subject for the fourth and fifth classes.[15] The academy also emphasised the teaching of

---

[12]   Thanks to Brian Cook, secretary of the Edinburgh Academical Club, who produced class figures for the period for us.
[13]   Magnusson 1974, p. 202.
[14]   Edinburgh Academy, Class Syllabuses, 1840s.
[15]   Magnusson 1974, pp. 63–4, 94, 112, 133–4; RLS, p. xxix.

Greek, and it is puzzling that Jenkin in later life was ignorant of the language.[16] According to syllabuses for the 1840s, Greek was taught from the third class, yet prizes in the subject were not awarded in Dr Cumming's class during the years of Jenkin's attendance, nor were Academical Club prizes in Greek for 1846 won by members of that class,[17] suggesting that Jenkin's year for some reason had omitted the study of Greek.

Charles Jenkin's claim that his son 'early distinguished himself amongst his class fellows' appears an exaggeration. At the end of his first year, Jenkin was placed eighteenth of the seventy-four pupils in the third class, with no subject prizes. In the fourth class, he rose to eleventh of seventy boys, though otherwise undistinguished. By the end of his final year at the school, his class position was sixth of fifty-four and he carried off prizes for best English scholar, best English verses and equal best Reciter. He was fourth geometrician, and mentioned in the mathematics list for the Edinburgh Academical Club Prize, which included boys from the sixth and seventh classes.[18] Considering Jenkin's later eminence as an engineer and scientist, this performance in mathematics appears mediocre. But the winner of the silver medal for geometry, and 'dux' of Dr Cumming's class for six consecutive years, was Jenkin's friend Peter Guthrie Tait (1831–1901). Tait graduated in the mathematical tripos from Cambridge as senior wrangler in 1852, the youngest on record, and his distinguished academic career included forty years in the chair of natural philosophy at Edinburgh, where he was again a contemporary of Jenkin.[19] Another outstanding mathematician in Dr Cumming's class was Allan Stewart, later involved in designing the Forth Bridge, and second geometrician in 1846.[20]

In the class above Jenkin was James Clerk Maxwell (1831–1879), the mathematician and physicist who developed the electro-magnetic theory of radiation and became first Professor of Experimental Physics at Cambridge University. Maxwell was to be a lifelong friend of Jenkin, and godfather to his youngest son, Bernard Maxwell Jenkin.[21] Although Maxwell's name is now better known than that of Tait, he was overshadowed by Tait first at school, then as a student at Cambridge, and later in competition for the chair of Natural

[16]   RLS, p. xliv; Jenkin wrote to his wife in 1864: 'I certainly should like to learn Greek, and I think it would be a capital pastime for the long winter evenings': RLS, p. lxxi.
[17]   Edinburgh Academy, Class Lists for 1844–46; Academical Club Prizes, 1846; Syllabuses, 1840s.
[18] The Academical Club prize adopted a new form in 1846, being awarded for the best overall performance across a number of written papers. The winner was Lewis Campbell, of the sixth class, later professor of Greek at St Andrews: Magnusson 1974, pp. 151, 163. Campbell was a friend of Jenkin in later life: see Chapter 6, below.
[19]   See the *Dictionary of National Biography*.
[20]   Magnusson 1974, p. 156.
[21]   Jenkin mss, notes by Anne Jenkin.

Philosophy at Edinburgh when both men were twenty-nine. Indeed, Maxwell had struggled in all subjects at the academy until he reached the mathematics class of James Gloag, where his potential was identified. From this time he achieved rapid progress across the curriculum. Gloag, a teacher at the Edinburgh Academy for forty years from its opening in 1824, had an outstanding record in nurturing mathematical talent. In the seven years to 1854, when the Senior Wranglership at Cambridge was considered the highest mathematical distinction in Britain, of Gloag's former pupils, one (Tait) had been Senior Wrangler, two (including Maxwell) second Wranglers, one ninth, one fourteenth and one thirtieth.[22] Jenkin later commended the teaching of mathematics at the academy as 'very good'.[23]

The presence of an inspired English teacher, Theodore Williams, nephew of the school's head, Rector Williams, was also of consequence for Jenkin. Besides his prizes for English in 1846, Jenkin was developing the love of drama and literature which was a passion throughout his life. In a scientist, this could be considered a serious distraction. Maxwell, like Jenkin, has been criticised for not sufficiently reigning in his intellect, and spending time on extraneous activities such as poetry and walking, rather than applying single-mindedly to science. Yet even Tait, the winner of glittering mathematical prizes, and by implication more closely focused than Maxwell or Jenkin, excelled in composing Latin verse, and in English and French. In any case science was not a subject directly offered at the academy, though Jenkin when in the Fifth Class organised a Philosophical Society at his home, where boys read papers on geology and chemistry, and set out on expeditions to collect geological and botanical specimens.[24] He afterwards said that his scientific education had begun at the Edinburgh Academy.[25]

Financial difficulties for the Jenkins intensified with a loss of investments in the West Indies,[26] and the father's failure to find employment. Charles Jenkin's final posting was to the *Myrmidon* in December 1845. In October 1846 he was praised for creditable conduct during 'a popular outbreak in Ireland', his promotion to commander was confirmed the following month and in December 1846 he received two commendations, one for 'zeal and activity', the other for gallantry during a fire.[27] Yet after his discharge the same month, Charles Jenkin did not serve again. His half-pay became the family's only income. The economies of life on the continent made a move abroad attractive, perhaps even essential. Thus Fleeming Jenkin's promising career at

---

[22] Magnusson 1974, p. 102.
[23] Samuelson committee, p. 122.
[24] Magnusson 1974, p. 154.
[25] Samuelson committee, p. 122.
[26] RLS, p. xxx.
[27] PRO, ADM 196/36, p. 693.

the academy was cut short, and his education took a new direction, particularly into modern languages and engineering. The Jenkins left Edinburgh in 1846, travelling via London to Germany. In Frankfurt-am-Main they lived in 'a pretty house outside the walls', and Jenkin attended a local school until the master died.[28] Here he learnt German. The family lived in Paris from 1847 until forced on by the insurrection of 1848. Jenkin's tutor was a Monsieur Deluc, with whom he continued his studies in French 'and took great delight in mathematics'.[29]

Paris in 1848 proved a formative experience for the fifteen year-old Jenkin. His mother, staunchly liberal in outlook, moved in radical circles which included political exiles from Italy. Jenkin's letters to an Edinburgh friend, Francis Scott,[30] reflect a boyish excitement with the course of events, and with his proximity to the action, and strong sympathies with the revolutionaries.[31] But events took an unpleasant turn. As it was considered unsafe for British citizens to stay in Paris, or even to take a direct route home, the Jenkins headed for Italy.

Though caught up in further episodes of civil unrest during their first year in Genoa, the family settled there for three years, from 1848 until 1851. They were evidently drawn by Mrs Jenkin's friendship with the Ruffinis, whom they had met during the latters' exile in Paris. Commander Jenkin, wrote his wife in September 1849, 'is more constant in his liking to Genoa than I have ever known him to any other place, but I begin now and then to hear something of return to England etc., at all events saving the casualty of war, we are to remain here another year at least – after that we shall I suppose begin our homeward movement.'[32] Charles Jenkin, unable to speak any language but English, and overshadowed by his sociable and brilliant wife, found continental life difficult, though the violent events of spring 1849 had given him an opportunity to act in defence of family and friends.

For a time, their son's education took priority. Through Augustine John Ruffini, then deputy for Genoa,[33] Fleeming Jenkin had been introduced to a school connected with the University, and he spent the first year in Italy preparing for university admission. The political changes made it possible for non-Catholics to attend there, and Fleeming Jenkin, known as 'Signor Flaminio' according to Stevenson, became the first Protestant student at Genoa.

---

[28] Jenkin mss, father's notes.

[29] Jenkin mss, father's notes.

[30] The son of Thomas Scott of Fettes Row: Edinburgh Academy archives.

[31] These are reproduced at length by RLS, pp. xxxi–xxxvii. It is not known whether the original letters have survived.

[32] Edinburgh University Library, Gen.1996/12/52, letter to Mrs Catherine Torrie, 10 September 1849.

[33] Soon after this Ruffini became head of the University. He was still in contact with Mrs Jenkin during the 1870s: RLS, pp. xxxi, xl, cxlvii.

This delighted his mother, as 'a living instance of the progress of liberal ideas'.[34] To gain admission, the sixteen year-old Jenkin had to produce certificates from the Edinburgh Academy and from M. Deluc in Paris, and pass two preliminary examinations. He needed to show competence in Latin, in order to follow lectures, and had to write essays in Latin and English. The oral trials in Horace, Tacitus and Cicero were, says Stevenson, 'much softened for the foreigner'.[35] No Greek was required, but three months into his course Jenkin had to take a test in Italian eloquence. In September 1849, a month before he went up to university, his mother wrote to a friend in Edinburgh that her son was 'growing very fast, a full half head taller than I am ... he is pursuing a course of surveying, and as the surveyors are always passing this vineyard he can refresh himself at leisure with the beautiful and abundant grapes'.[36] At the university, said his father, Jenkin 'applied himself to all the Instruction there, Mathematics, Graphics, Rudiments of Engineering, Physics, and he acquired great facility in the Italian Language'.[37]

Bancalari, the professor of natural philosophy, had at Genoa the best mounted physical laboratory in Italy, and was particularly interested in electromagnetism. This was the main subject of Jenkin's study, and after a year – questioned in Latin and answering in Italian – he obtained his MA with first class honours. He was later to describe his scientific education under Professor Bancalari as 'fairly good'.[38] All the while his mother had been teaching him to play the piano – she was a renowned performer of Chopin and Schubert – and continuing his tuition in drawing, for which he was awarded a silver medal from the art school.[39] 'Fleeming has always shown a turn for drawing', said his father.[40] Henrietta was a naturally talented though untrained artist, and the captain had also dabbled in water-colours to pass the time on St Helena, though Stevenson is not complimentary about his artistic ability.[41] Jenkin's own skill is demonstrated by the quality of sketches of his mother, of himself and of his children, reproduced in Stevenson's *Memoir*.[42]

After his graduation in 1850, the Jenkins did not immediately leave Italy. Fleeming Jenkin's first employment was in an engineering shop in Genoa under Philip Taylor of Marseilles (1786–1870). One of Taylor's sons was 'chief friend and companion' of Jenkin at this time, and Jenkin apparently

[34] RLS, p. xl.
[35] RLS p. xliv.
[36] Edinburgh University Library, Gen.1996/12/52, letter to Mrs Catherine Torrie, 10 September 1849.
[37] Jenkin mss, father's notes.
[38] Samuelson committee, p. 122.
[39] RLS, p. xlv; *The Scotsman*, 9 February 1885.
[40] Jenkin mss, father's notes.
[41] RLS, pp. xix, xxiii.
[42] Some of which are reproduced in this biography.

enjoyed his time there: 'I began my engineer's career in [Taylor's] establishment and he was very kind to me at Genoa'.[43] Taylor, a partner in the London engineering company of Taylor and Martineau which was dissolved in about 1827, had afterwards established a business in Marseilles as ironfounder and marine engine manufacturer with substantial contracts for the French and Italian governments. His main business was in supplying machinery to several local steam navigation companies. His branch at St Pietro d'Arena in Genoa dealt with silk mill contracts, though in another account the Italian establishment is described as a locomotive shop.[44]

By the time Jenkin went to work for Taylor, the Marseilles business had been sold to an association known as the Compagnie des Forges et Chantiers de la Mediterranee. Then, in 1851, Taylor disposed of the Genoa works and retired to Marseilles. His withdrawal from business happened at about the time that his son, Jenkin's friend, was accidentally drowned.[45] These circumstances, along with the death of Aunt Anna in Genoa in January 1851, led the Jenkin family finally to return to England. 'After much consideration', recalled his father,' it was thought best to visit Manchester where it was agreed with Mr William Fairbairn senior to receive Fleeming Jenkin as a Pupil to learn Engineering under his care. A handsome premium was paid and we took a House in Ardwick near for the convenience of all.'[46] The house was 1 Polygon Avenue, and Jenkin's three-year pupilage was with the renowned civil and mechanical engineer William Fairbairn, a man at the top of his profession.[47]

Jenkin's work was similar to that which he had done with Taylor in Genoa, working from half past eight until six, 'filing and chipping vigorously in a moleskin suit'.[48] His time was spent in the workshops, or the drawing office, and sometimes out on site, for example 'to regulate a governor at the large mill at Saltaire'.[49] Despite his previous experience with Taylor and a scientific education which gave him a very great advantage over most fellow

---

[43] Kelvin, Glasgow, J24, Jenkin to Thomson, 24 September 1859.
[44] See *The Engineer*, XXX (2), 8 July 1870, pp. 20 and 25. Taylor was described in this death notice as Civil Engineer, Knight of the Legion of Honour in France, and of the Order of St. Maurice in Italy. Keith Moore of the Institution of Mechanical Engineers supplied this reference.
[45] *The Engineer*, XXX (2), 8 July 1870, p. 20; Glasgow, Kelvin, J24, Jenkin to Thomson, 24 September 1859.
[46] Jenkin mss, father's notes.
[47] Polygon Avenue still exists, a mile south of Piccadilly and about the same distance from Fairbairn's works in Ancoats, although the house in which the Jenkins lived no longer stands. For this information we are grateful to John Cantrell.
[48] RLS, pp. xlvii.
[49] Kelvin, Glasgow, J26, Jenkin to Thomson, 5 January 1860. The building of Saltaire Mills, near Bradford, was Fairbairn's last major commission before his retirement in 1853: R.A. Hayward, 'Fairbairn's of Manchester: the history of a nineteenth-century engineering works', unpublished MSc. thesis, UMIST, 1971, pp. 2.10–2.12.

pupils, Jenkin later admitted that during the first six months with Fairbairn he 'felt very awkward', for he had compared himself unfavourably with a slightly older man, a graduate of King's College who had been as well prepared as Jenkin himself. Yet by the time he left Manchester, Jenkin had become 'a fair average workman' in comparison with shop-floor apprentices of the same age who had several years' start on him. He had been put in charge of the pattern shop for a time, and reckoned that he was capable of building a locomotive by the end of his pupilage.[50]

This new life under 'the dim skies and [in] the foul ways of Manchester'[51] did not lack compensations. Jenkin never saw his shop-floor duties as unremitting drudgery, nor in any way beneath him. Stevenson emphasises that 'anything done well, any craft, despatch or finish, delighted and inspired him'.

> Nothing indeed annoyed Fleeming more than the attempt to separate the fine arts from the arts of handicraft; any definition or theory that failed to bring these two together ... had missed the point; and the essence of the pleasure received lay in seeing things well done. Other qualities must be added ... but this, of perfect craft, was at the bottom of all.[52]

But to Jenkin, the work of an engineer amounted to more than mere craft. Stevenson, who had narrowly escaped a career in engineering and retained a hearty dislike of the activity, described his mentor's fascination with the machine, 'in which iron, water and fire are made to serve as slaves, now with a tread more powerful than an elephant's, and now with a touch more precise and dainty than a pianist's'. Jenkin showed 'a certain bitter pity' for the weakness which Stevenson displayed in not sharing a taste for machinery.

> Once when I had proved, for the hundredth time, the depth of this defect, he looked at me askance: 'And the best of the joke,' said he, 'is that he thinks himself quite a poet.' For to him the struggle of the engineer against brute forces and with inert allies was nobly poetic.[53]

Jenkin's middle son, Frewen, shared this single-mindedness and passion for engineering. After his father's death, Frewen Jenkin wrote to Stevenson:

> In one thing I am better off than you for I love engineering & [my father and I] seemed to feel exactly alike about it. I remember when I brought down the first screw I ever made to shew him in the study, he looked at it & smiled & said

---

50 Samuelson committee, p. 122.
51 RLS, p. xlvii.
52 RLS, p. xlviii.
53 RLS, p. xlix.

'does it not make you feel good all over', which is exactly what it did, and why I had brought it to him.[54]

Jenkin's respect for craft skills, and for the artisans he first encountered during his training in Manchester, was pivotal to his later interest in technical education and trades union issues. More than that, he came to believe that shop-floor experience was essential to any engineer with real ambition:

> If a man is ambitious and means to distinguish himself as an Engineer. If he means to rise by sheer ability into the front rank. If he means not to be a mere inspecting engineer and man of business (like many of the best paid men) but a real designer able to carry out new and difficult work; then he must work as a workman, and if he has the kind of ability and common sense which I have in my mind the temptations he will encounter and the course associates he will meet will do him no harm. As a rule the young men who go as apprentices to Mechanical Workshops are not Gentlemen. My meaning in fine is that if a man is strong mentally and morally and ambitious to boot he must work with his hands. A young man of merely good average ability and character need not.[55]

Although the Jenkins' move to Manchester had been expressly to further Jenkin's career, the family had prior acquaintance in that city. Writing from Genoa in 1849, Jenkin's mother had referred to 'the dear Bells' of Manchester.[56] It is likely that these were originally Edinburgh acquaintances, for George Joseph Bell (1770–1843) had been professor of Scots law in the University of Edinburgh; his brother, Sir Charles Bell (1774–1842) was a distinguished anatomist specialising in nervous diseases.[57] Jenkin continued his studies in Manchester, three evenings a week, with Dr Bell, son of George Joseph. Dr Bell was interested in the geometry of Greek architectural proportions, a subject which appealed to Jenkin as it combined art with science, and from which Stevenson believed Jenkin derived his long-standing appreciation of 'things Greek'.[58]

---

[54] Yale, Beinecke, 4982, C.F. Jenkin to Stevenson, 9 August 1887.

[55] University of St Andrews Library, Forbes correspondence (in) 1865–62, Jenkin to Thomson, 5 July 1865.

[56] Edinburgh University Library, Gen.1996/12/52, H.C. Jenkin to Mrs Catherine Torrie, 10 September 1849. The friendship with the Bells continued, and Miss Bell of Manchester was among visitors to Jenkin's home at Claypole ten years later: RLS, p. lxix.

[57] See the *Dictionary of National Biography*; also Chapter 5 below re Jenkin's paper on Mrs Siddons, written from notes provided by Professor George Joseph Bell.

[58] RLS, p. l. A Joseph Bell who attended the Edinburgh Academy after Jenkin had left, became the model for Conan Doyle's Sherlock Holmes as Dr Bell (Magnusson 1974, p. 208), but no evidence has been found of a relationship to the Dr Bell in Manchester.

Jenkin acquired a notable acquaintance during his stay in Manchester when he became an intimate in the family of Mrs Gaskell. How he came to meet the novelist is not clear. The Gaskells were friends of Fairbairn, who may have been the means of Jenkin's introduction to their house, Plymouth Grove.[59] His Saturday afternoon calls there were a highlight of Jenkin's week for almost four years.[60] He tried out ideas upon Mrs Gaskell, but the main feature of his visits was noisy and animated argument with the Misses Gaskell. 'Fleeming came to tea; and talked incessantly and very cleverly ...'.[61] A less easy relationship between Jenkin and Mr Gaskell is hinted in a letter from Mrs Gaskell to her daughter Marianne: 'I asked [Fleeming] to tea on Sunday – which I thought was a bold stroke, but however he agreed to come ... For a wonder Papa and Fleeming talked to each other a good deal'.[62] The Gaskells remained close friends of Jenkin and of his mother, to the extent that Jenkin, during a busy period of his life in 1862, spent time vetting prospective London lodgings for Mrs Gaskell, sending 'a memorandum full of details' to help her make a choice.[63]

Manchester in the 1850s was a cold and depressing contrast with Genoa, and Jenkin's work 'infernally dirty'.[64] Worse, the family's means were further stretched by the expense of life in England. According to Stevenson, the Jenkins 'did not practise frugality, only lamented that it should be needful'; and while Mrs Jenkin constantly complained of 'dreadful bills', she was 'always a good deal dressed'. Pupil engineers were not paid during their training, and because of Fairbairn's standing the premium for Jenkin's pupilage was high at £300.[65] There is other evidence to confirm the family's straitened finances while in Manchester, apart from Stevenson's description of Jenkin's lack of overcoat and use of old newspapers to keep warm on railway journeys. Charles Jenkin made renewed efforts to find work. In January 1852 he applied to the Board of Admiralty for appointment as a mail agent on board contract vessels

[59] Chapple and Pollard (eds), 1966, p. xvii.
[60] Chapple and Pollard (eds), 1966, pp. 363–64, no. 259, Gaskell to Marianne Gaskell, 27 July 1855.
[61] Chapple and Pollard (eds), 1966, p. 790 no. 602, Gaskell to Marianne Gaskell, undated.
[62] Chapple and Pollard (eds), 1966, pp. 363–64, no. 259, Gaskell to Marianne Gaskell, 27 July 1855.
[63] Chapple and Pollard (eds), 1966, pp. 684–85, no. 505, Gaskell to Marianne Gaskell, 1/2 May 1862.
[64] RLS, p. xlvii.
[65] The usual rate for a pupilage at this time was £50 to £100, although a five-year apprenticeship could cost as much as £1000. In 1868, Jenkin said that 300 to 500 guineas was usual for a three year pupilage with a top rank engineer; John Penn's pupils had paid £400 during the 1850s: Samuelson committee, pp. 133–38. For background on premium apprenticeships in engineering, see Charles More, *Skill and the English Working Class, 1870–1914* (London: Croom Helm, 1980), pp. 104–6.

to Australia. This was followed in March by a letter to the Earl of Hardwicke, an old friend who was then Postmaster General, suggesting that seaman gunners should be deployed aboard these vessels to protect against pirate attacks, and asking for Hardwicke's support for his application.[66] Hardwicke referred the letter to the Admiralty, but nothing came of it.

It was shortage of money which led Henrietta Camilla Jenkin to write the novels which earned her own entry in the *Dictionary of National Biography*. Mrs Jenkin is said to have had no natural aptitude for literature, but turned to writing 'under pressure of poverty'.[67] Her first novel, *Violet Bank and its Inmates*, published in 1858, was not successful. Dissatisfied with the publisher, she asked for an introduction to Elizabeth Gaskell's own publisher, George Smith,. Mrs Gaskell wrote to Smith, about Mrs Jenkin's new novel, *Cousin Stella*: 'It is not her first work of fiction – she wrote *Violet Bank* about a year ago, which Messrs Hurst and Blackett published, and which was well-reviewed. She thinks however that Messrs. H. and B. behaved shabbily to her ... and is anxious to have an introduction to you, for this second West Indian novel; which I have not seen, but of which Signor Ruffini ('Doctor Antonio') thinks very highly.'[68] Whether impressed with the novel, or in deference to his best-selling client, Smith immediately accepted *Cousin Stella* and sent a cheque to Mrs Jenkin.[69] When the book was published, Mrs Gaskell responded with delight to her publisher: 'You never, no, *never*, sent a more acceptable present than *Cousin Stella* ...'[70] Henrietta Jenkin had a further half dozen novels published between 1861 and 1874, along with at least one story written in French, though none met with the success of *Cousin Stella*.[71]

At the end of Jenkin's apprenticeship with Fairbairn, he was engaged by George Willoughby Hemans (1814–85) as a surveyor on the construction of the Lukmanier Railway in Switzerland. Hemans was a civil engineer whose best known work was on railways and bridges, and Jenkin worked for him for six months.[72]

---

[66] British Library, Add. Mss 35789 f.163. Charles Philip Yorke, 4th Earl of Hardwicke, (1799–1873), served briefly in Lord Derby's cabinet. His acquaintance with Charles Jenkin presumably dated from his attendance at the Royal Naval College, Portsmouth, between 1813 and 1815: *Dictionary of National Biography*.
[67] *Dictionary of National Biography*; Stevenson concurs with this view: RLS, p. xxiv.
[68] Chapple and Pollard (eds) 1966, p. 527, no.412, Gaskell to George Smith, 10 February 1859.
[69] Harvard University, Houghton Library for Rare Books and Manuscripts, Fleeming Jenkin to Elizabeth Gaskell, 7 March 1859.
[70] Chapple and Pollard (eds) 1966, p. 562 no.434, Gaskell to George Smith, 29 June 1859.
[71] *The Scotsman*, 9 February 1885.
[72] Obituary of George Willoughby Hemans, *Institution of Civil Engineers Proceedings*, LXXXV (1886), pp. 394–99; Obituary of Fleeming Jenkin, *Institution of Civil Engineers Proceedings*, LXXXII (1885), p. 365.

In July 1855, Elizabeth Gaskell wrote to her daughter Marianne that Jenkin was not to return to Manchester. 'Fleeming ... has got a situation at Penn's, Greenwich; some place as known near London, he says, as Fairbairn's here. He came down for three days to wind up his Manchester affairs entirely..' His mother had gone abroad for six months. 'He himself means to stay at Penn's for nine months or a year, and then get some engagement [in] Canada or Australia'.[73] By his father's account, Jenkin had gone to John Penn, who made marine engines for warships, with a letter of introduction from William Fairbairn, and though there was no vacancy, Penn managed to accommodate the young man as a draughtsman in his 'first rate establishment'.[74] John Penn, FRS, (1805–78), who had joined the Institution of Civil Engineers in 1826, moved among the highest ranks of engineers. His application for full membership of the Institution in 1845 had been supported by, among others, Marc Brunel, Bidder, Farey, Cubitt and Joseph Maudslay.[75] When Jenkin, by then a professor and himself FRS, was elected a member of the Institution of Mechanical Engineers in 1875, John Penn was one of his three sponsors.[76]

Jenkin's work was hard, the hours longer than with Fairbairn, for the firm was busy finishing engines for new gunboats and frigates to be used in the Crimean war. Working twelve or thirteen hours a day, Jenkin, according to Stevenson, was surrounded by 'uncongenial comrades' and staying in lodgings nearby, without the company of his mother[77] – though Mrs Jenkin did live in Greenwich, at 8 Blackheath Terrace, during at least a part of his time with Penn.[78] Jenkin again found solace in his work, and wrote, in a final letter to Frank Scott in Edinburgh: 'Luckily I am fond of my profession, or I could not stand this life'.[79] He visited friends in town on Sundays, 'and seem to swim in clearer water, but the dirty green seems all the dirtier when I get back'. Otherwise he managed to continue his studies in engineering and mathematics, read Carlyle and poetry, took singing lessons and maintained a circle of female correspondents.[80]

Despite the tribulations of working in Greenwich, Jenkin did fulfil his intention of staying a year with Penn. But he never carried out the plan to emigrate to the colonies, perhaps because of a significant change in his personal life. On his move to London in 1855, Jenkin had taken a letter of

[73] Chapple and Pollard (eds) 1966, pp. 363–64, no. 259, Gaskell to Marianne Gaskell, 27 July 1855; p. 357, no. 253, Gaskell to Marianne Gaskell, July 1855.
[74] Jenkin mss, father's notes.
[75] Institution of Civil Engineers, membership records.
[76] Institution of Mechanical Engineers, membership records.
[77] RLS, p. lii.
[78] Chapple and Pollard (eds) 1966, p. 809 no. 637, Gaskell to William Chambers, 29 February [1859?].
[79] RLS, p. liii.
[80] RLS, pp. lii–liii.

introduction from Mrs Gaskell to Alfred Austin.[81] Austin, one of a distinguished family of lawyers originally from Creeting Mill in Suffolk, lived at Sussex Place, Regents Park, with his wife, *née* Eliza Barron, and their only child, Anne.[82] It is not certain whether this was Jenkin's first meeting with Anne, for she had been a close friend of the Gaskell daughters before 1851, and the girls often stayed with each other. Anne was at Plymouth Grove several times during Jenkin's time in Manchester, including a visit after Christmas 1851.[83]

Alfred Austin, a barrister like his brothers, had turned to public service. He carried out a Commission of Inquiry into the state of the poor in Dorset, then became a Poor Law Inspector in Worcester, moving to Manchester to tackle the consequences of the mass influx of Irish immigrants during the potato famine of the 1840s.[84] It was probably at this time that his daughter met the Gaskells. After the Austins' move back to London, where Alfred Austin dealt with the aftermath of a cholera epidemic, he was appointed Permanent Secretary to Her Majesty's Office of Works and Public Buildings. After retiring in 1868, he was awarded a civil Companionship of the Bath in recognition of his work.[85]

Fleeming Jenkin struck an instant rapport with Alfred and Eliza Austin, though his relationship with Anne Austin took longer to develop. Stevenson describes the Austins as 'a centre of much intellectual society', who counted John Stuart Mill and Thomas Carlyle among their wide acquaintance.[86]

> When Fleeming presented his letter he fell in love at first sight with Mrs Austin and the life and atmosphere of the house. There was in the society of the Austins, outward, stoical conformers to the world, something gravely suggestive of essential eccentricity, something unpretentiously breathing of intellectual effort, that could not fail to hit the fancy of this hot-brained boy ... Here he found persons who were the equals of his

---

[81] Jenkin mss, notes of Anne Jenkin.

[82] London directories show the Austins at 13 Sussex Place in 1855 and 1858. They had moved to Spring Grove, Isleworth, by 1860: Glasgow, Kelvin, J41, Jenkin to Thomson, 7 September 1860. The careers of Alfred Austin's older brothers, John (1790–1859) and Charles (1799–1874), are described in the *Dictionary of National Biography*. John Austin lived in Paris between 1844 and 1848 but there is no evidence that he knew the Jenkin family there. See Chapter 6, below.

[83] See for example various letters from Mrs Gaskell to her daughter during the early and mid 1850s: Chapple and Pollard (eds) 1966, p. 144 no. 90; p. 146, no. 92; p. 170, no. 107; p. 831 no. 91a; also Barbara Brill, 'Annie A. and Fleeming', *Newsletter of the Gaskell Society*, 1 (March 1986), for which reference we are grateful to Elaine Greig of the Writers' Museum, Edinburgh.

[84] RLS, p. lvi.

[85] RLS, p. lvi; the Jenkin mss contain a copy of Alfred Austin's letter to Gladstone accepting the honour, 24 September 1869.

[86] John Stuart Mill, *Autobiography* (Oxford, 1971).

mother and himself in intellect and width of interest, and the equals of his father in mild urbanity of disposition.[87]

Jenkin saw the Austins' marriage as an ideal model for his own future happiness, and after a time realised that their daughter could be his perfect partner. More than two years after their first meeting, 'this boyish-sized, boyish-mannered and superlatively ill-dressed young engineer', Figure 2.4, asked to be allowed to pay his addresses to Miss Austin,[88] Figure 2.5. The parents, approving his character, did not object to his lack of means. Anne Austin's initial indifference was worn down by Jenkin's devotion and sincerity, helped by the fact that her mother enthusiastically supported his cause. From

Figure 2.4: Self-portrait of Fleeming Jenkin, aged twenty-six (Jenkin mss)

Figure 2.5: Anne Jenkin, from a pencil drawing by Fleeming Jenkin, undated (Jenkin mss)

this time, Jenkin's relationship with Anne was the bedrock upon which he built his life. She was his intellectual equal, and brought him into contact with new spheres of learning and activity.

Anne Austin had been, in the words of Stevenson, 'brought up ... like her mother before her, to the standard of a man's acquirements'.[89] Yet though allowed to study Greek, this fact was kept secret, 'like a piece of guilt'; furthermore, even these liberal parents would not tolerate a girl learning the violin. An Edinburgh friend, Mrs Henry Roscoe, echoes Stevenson: 'She had had what was called in those days a 'boy's education' i.e. she had been taught Latin and Greek (not mathematics I think) and I know that Professor Sellar

---

87   RLS, pp. lviii.
88   RLS, pp. lix–lx.
89   RLS, p. lvii.

admired her work ...'[90] Intellectual curiosity remained a hallmark of Anne Austin. Mrs Roscoe recalled having first met her at a Latin class in Edinburgh run for women and part-time students by William Young Sellar. While Fleeming and Anne Jenkin were both renowned as 'brilliant talkers' in fact it was she who was 'the more eloquent, and in a way more concerned with ideas'.[91] Anne was 'more thoroughly educated than he. In many ways she was able to teach him, and he proud to be taught; in many ways she outshone him, and he delighted to be outshone'.[92] She was also a gifted actress and her views on drama strongly influenced those of Jenkin.[93]

Henrietta Jenkin, the strong-willed mother who had supervised her only child's education and taken a close interest in every aspect of his life before he knew Anne, stayed close to him. During the dark days in Greenwich, he had written to a friend of his visits to Henrietta in London, listening to 'mamma's projects *de* Stowting. Everything turns to gold at her touch – she's a fairy and no mistake ... Even you don't know half how good mamma is ... She teaches me how it is not necessary to be very rich to do much good. I begin to understand that mamma would find useful occupation and create beauty at the bottom of a volcano. She has little weaknesses, but is a real, generous-hearted woman, which I suppose is the finest thing in the world'.[94] Anne Jenkin later wrote about Henrietta that 'generosity in every sense was her greatest quality – quickly she knew that Fleeming cared for me ... she never *once* grudged me Fleeming's affection'.[95]

Fleeming Jenkin's marriage to Anne Austin, 'a pretty walking country wedding',[96] took place in February 1859, at Northiam, Sussex, where Charles Jenkin had spent part of his childhood, and where the Austins also had connections. Jenkin wrote a joyful letter, to which 'Annie' added her name, to Mrs Gaskell following his wedding:

> If you please we want to be congratulated, told that we are charming people, quite sure to be happy, worthy of one another and as many pleasant truths as you can think of ... tell us that our wedding cake is

---

[90] Jenkin mss, Mrs Roscoe's recollections of Anne Jenkin.
[91] Jenkin mss, Mrs Roscoe's recollections of Anne Jenkin.
[92] RLS, p. lxvi.
[93] See for example Booth and Mehew (eds) 1995, V, p. 61 and n., Stevenson to Henley, 2 January 1885. This point is discussed further in chapter 6, below.
[94] RLS, p. liii–liv.
[95] Yale, Beinecke, 4979, Anne Jenkin to RLS, 2 September [1886?].
[96] Chapple and Pollard (eds) 1966, p. 538 no. 418, Gaskell to Charles Eliot Norton, 9 March 1859.

admirable that our little country wedding is the *beau ideal* of weddings and we shall be so pleased and believe every word of it ...[97]

After this there was no more mention of Canada or Australia, for along with his domestic contentment, Jenkin had found a new and challenging *métier* in electrical engineering, in the new world of submarine telegraph cables.

[97]Harvard University, Houghton Library for Rare Books and Manuscripts, Fleeming Jenkin to Elizabeth Gaskell, also signed by Annie Jenkin, 7 March 1859.

Chapter 3

Employed about the Ocean Cables[1]

After a honeymoon of two days in Oxford, Mr and Mrs Jenkin went to Birkenhead, where four ocean-going ships were decked with flags in tribute to the bride.[2] Fleeming Jenkin, still only twenty five years old, had become a person of some consequence in the Birkenhead telegraph works of R.S. Newall and Company.

Jenkin had moved from John Penn's in 1856, to work as a draughtsman on railway projects for Liddell and Gordon in London.[3] Liddell and Gordon's main business was in civil engineering, but they were also partners with Newall in his rope- and cable-making business, established in 1840 in Gateshead.[4] With these complementary interests, able to lay as well as manufacture cables, the partners were well placed in the new and rapidly expanding field of undersea telegraphs. It was in 1857 that Fleeming Jenkin left railway work, moving into the employment of R.S. Newall and Company and the technological challenges of submarine telegraphy. Much of Jenkin's time between 1857 and 1861 was spent at Newall's factory in Birkenhead, or away on expeditions to lay or retrieve undersea cables in the Mediterranean.[5]

Charles Liddell (1813–94), known mainly as a railway engineer, became involved in submarine telegraphy in 1855, when he laid the Varna and Balaklava cables during the Crimean War.[6] His partner Lewis Dunbar Brodie Gordon (1815–76), previously an associate of Marc Brunel on the Thames tunnel, was noted as a designer and manufacturer of wire ropes. Gordon built his reputation upon an idea which he had picked up at the Freiberg School of

---

[1]   Chapple and Pollard (eds) 1966, p. 538 no. 418, Gaskell to Norton, 9 March 1859. Mrs Gaskell described Jenkin as 'a young engineer, employed about the ocean cables ...'
[2]   Chapple and Pollard (eds) 1966, p. 538 no. 418, Gaskell to Norton, 9 March 1859; RLS, p. lxiv.
[3]   Jenkin's application to join the Institution of Civil Engineers in 1859 stated that he had worked for three years for Liddell and Gordon in Birkenhead and London. His obituary in the Institution's *Proceedings* mentions railway engineering with Liddell and Gordon: LXXXII (1885), p. 365.
[4]   F.W.D. Manders, *A History of Gateshead*, (Gateshead, 1973), p. 79.
[5]   In 1867 Jenkin was vague about the date he had started work on submarine telegraphy, putting it at 1857 or 1858: National Maritime Museum, Ms 88/078. However a letter he wrote to Anne Austin in April 1858 makes it clear that he had been working in the field during the autumn of 1857: RLS, p. lxxv.
[6]   *The Engineer*, 24 August 1894.

Mines in 1838. A distinguished German engineer, Wilhelm August Julius Albert, had developed a rope made from untwisted wire, for use in mining. The design, freely communicated to Gordon, never profited Albert. When Gordon returned to Britain – he was appointed to the first British chair of engineering, the Regius Chair of Civil Engineering and Mechanics at Glasgow in 1840[7], at the age of twenty-four – he took out a patent jointly with Liddell and Robert Stirling Newall for an improved version of this rope. R.S. Newall and Co was established in 1840 and produced wire ropes on the new principle for various purposes, including the rigging of ships. After ten successful years, the trio sent a silver vase to Albert's widow, a tribute to the inventor, inscribed with the miners' salutation 'Gluck auf!'. Frau Albert's response is not recorded.[8] Gordon abandoned his academic career in 1855, apparently after encountering problems with jealous colleagues and obstructions in introducing practical elements to his teaching.[9] It has been suggested that he was unable to continue with both academic and commercial work,[10] though the practice of engineering professors carrying a heavy consultancy load was then becoming widely accepted.[11] Most likely Gordon saw a promise of more lucrative and substantial work with Newall, and chose to leave the petty squabbles of academia behind him. By chance this led to personal misfortune, for Gordon's health was ruined after a shipwreck on his way home from the Red Sea cable expedition of 1859, and although he lived until 1876 and remained a partner in the company, he was never again able to work.

In 1850, an attempt by Jacob and John Watkins Brett[12] to establish a cross-channel telegraph had demonstrated that, while the idea of transmitting electrical currents beneath the sea was feasible, the core needed more substantial protection. The first cable was broken by chafing on rocks after just one day. Newall, who had become interested in telegraphy in 1848, had already conceived and patented the idea of sheathing the core of the cable – which was formed from twisted copper wire insulated with gutta percha – with wire rope.[13] He took out injunctions against two rope-makers who were trying to produce a similar cable to his own for a renewed attempt on the cross-channel route. As the concession on that line was shortly to lapse, the Submarine Telegraph Company had no option but to employ Newall to make and lay a second cable, in place of the Gutta Percha Company.[14] This was successfully completed in

---

[7]   Smith and Wise 1989, p. 654.
[8]   This story, and the account of Gordon's life and career, is told in Constable 1877, especially pp. 44–9.
[9]   Birse 1983, p. 64.
[10] Constable 1877, p. 45.
[11] See Chapter 5, below.
[12] See *Electrical Trades Directory*, 1898, p. 75.
[13] Newall 1882, p. 2.
[14] Smith 1974 [1891], Ch. III.

1851 and proved to be a significant breakthrough in undersea telegraphy.[15] Newall had already set a pattern with his proceedings to protect various patents and to claim credit for his innovations. This taste for constant litigation and controversy ultimately rebounded, for his main competitors, Glass, Elliot and Co., achieved a near monopoly with the deep-sea telegraph companies in the early 1860s. R.S. Newall eventually gave up cable-making and cable-laying, turning to chemical manufacture and later asbestos processing.[16] During the 1850s, though, he and his partners were a major force – in fact they saw themselves as the technological leaders[17] – in armouring and laying deep-sea cables. In 1857, when Fleeming Jenkin arrived in Birkenhead, they were completing a commission to make half of the cable for the first attempt to span the Atlantic.

Gordon had been casting around for a young engineer who could undertake experimental work as well as more practical duties. William Thomson, professor of Natural Philosophy in the University of Glasgow, had convinced Gordon that 'scientific deductions from established principles' could be a more reliable and economical method of advancing submarine telegraphy, than expensive practical trials from which even highly reputable engineers such as Werner Siemens had drawn mistaken inferences.[18] Thomson saw the future in a harmony of theory and practice.[19] In July, 1857, Gordon wrote to Thomson about his need for an employee who understood scientific principles:

> We are sadly at a loss for a Philosophic assistant who is at the same time a practical man; or who could readily become one. We could afford to pay him well and to give him large scope for experiments. Can you help us? If you would like us to supply you with means for experiments we should be very happy to supply them on the natural principle of the industrial results being for our advantage. But Newall would like to have a skilful experimenter and suggester beside him and if you know an aspirant philosopher and experimentalist let us hear of him.[20]

This was evidently the vacancy which Jenkin filled soon afterwards, for Gordon found the 'young man of remarkable ability' already in his and Liddell's employment.

---

[15]   It was still in service thirty years later: National Maritime Museum, TCM/7/2.

[16]   Newall and his partners gradually withdrew from cable-making after their firm went public in 1870, becoming the National Telegraph Manufacturing Co. Ltd. with a capital of £400,000. See also C.A. Russell, 'Robert Stirling Newall, 1812–89, wire rope inventor and chemical manufacturer', in Jeremy (ed.) 1985, IV, pp. 433–5.

[17]   Galton committee, p. 258.

[18]   See Smith and Wise 1989, pp. 662–6. For details of the life and work of Thomson, later Lord Kelvin, (1824–1907), see Smith and Wise 1989;   also Sloan 1996.

[19]   Smith and Wise 1989, p. 671.

[20]   Kelvin, Cambridge, G136, Gordon to Thomson, 3 July 1857.

Newall and Co. had been responsible for a string of submarine telegraphs after their cross-channel triumph – connecting Britain with various points on the Continent, and with Ireland; bringing offshore islands of Canada into communication with each other and with the mainland; providing a temporary link across the Black Sea for military purposes during the Crimean War.[21] When the contract was awarded to construct the first Atlantic cable in 1857, none of the leading cable-makers – all of whom were British – had the capacity to produce the whole 2500 miles required in time for the short summer cable-laying season, so the work was shared between Newall and Co. and Kuper and Co., shortly afterwards renamed Glass, Elliot and Co., Figure 3.1, and increasingly Newall's rival. A bitter row ensued when the cable was completed and it was discovered that the halves had been made with wires twisting in

Figure 3.1: Glass Elliot Works (*Illustrated London News*, 14 March 1857)

opposite directions. Although Newall produced a convincing defence, that his firm had worked from a sample supplied by the Atlantic Telegraph Company,[22] his competitors Glass, Elliot had close links with the telegraph company's directors, and the effect of the dispute was to further marginalize Newall from the Atlantic project. In the event it proved possible to splice the two portions of cable, and the expedition proceeded as planned without further involvement by Newall and Co. The failure of this first attempt to cross the Atlantic and of a second effort the following year showed up deficiencies in both the technology

---

[21]   See the *Gateshead Observer*, 21 August 1852;   13 January 1855.
[22]   *Gateshead Observer*, 1 August 1857;   Galton committee, p. 254.

and the understanding of electricity. A variety of problems relating to the manufacture, laying and operation of cables – and from diverse fields including mechanical, civil and electrical engineering, as well as navigation and oceanography – remained to be solved before deep-sea telegraphs could be laid and operated with confidence of complete success.

Most of the successful submarine cables working at this time were in shallow waters, of 150 fathoms or much less. The exception was one laid in 1854 from La Spezia in northern Italy to Corsica, a distance of seventy nautical miles and to a depth of 325 fathoms, by the Brett brothers, a former Newall employee called John Thompson,[23] and Henry Vernon Physick,[24] on behalf of the manufacturers, Kuper and Co., for the Mediterranean Telegraph Company.[25] This line worked without interruption for ten years. In 1855, the Bretts tried to complete the link from Europe to North Africa with a cable from Sardinia to Algeria, a distance of 130 miles. The depth turned out to be more than double the estimate of 800 fathoms. After two failures, the expedition ran short of cable a few miles from Galite Island, off the Tunisian coast, and the third attempt was abandoned. Newall and Liddell, whose reputation had been enhanced by their rapid installation of temporary unarmoured cables in the Crimea in 1855, were employed by the Mediterranean Telegraph Company in 1857 for another attempt on the line in the autumn of that year. It was Jenkin who fitted out a large steamer, the *Elba*, for what was to prove a successful expedition.[26]

Though Newall and Co. were based in Gateshead,[27] most of their work on submarine telegraphs was carried out elsewhere.[28] The premises in Cathcart Street, Birkenhead, had been taken late in 1856 to make the Atlantic cable, and this factory remained open until at least 1861 as further cable contracts followed.[29] It was situated in a warehouse next to the Great Float, where the core, supplied by the Gutta Percha Company, was armoured at a rate of more

---

[23] Smith 1974 [1891], p. 31.

[24] Also known later in his career as Henry Vernon (d.1894): see *Electrical Trades Directory*, 1895, pp. 102–3.

[25] National Maritime Museum, TCM/7/2.

[26] RLS, p. lxxv.

[27] Robert Stirling Newall was active in Newcastle and Gateshead affairs throughout his life, and was twice Mayor of Gateshead: see Russell in Jeremy (ed.) 1985, IV, pp. 433–5.

[28] For example the cross-channel cable had been completed at the Submarine Telegraph Company's works in Wapping: *Gateshead Observer*, 20 September 1851. There had also been a factory in Sunderland, which had proved to be much too small: Smith 1974 [1891], pp. 20–21.

[29] The firm also maintained an office in Liverpool from *c*.1855 to 1876. For this information, drawn from Birkenhead and Liverpool directories between 1855 and 1876, we are grateful to David Thompson, archivist, Metropolitan Borough of Wirral. It is possible that Newall was in Birkenhead and Liverpool for a longer period than the dates given.

than one hundred miles a week. Including the wire-drawers' works, seven hundred men were employed.

> In the outer factory the gutta percha covered wire, when received, in two mile lengths, is carefully tested by the galvanometer, for the purpose of insuring complete insulation; and, after being covered with yarn, it is saturated with a mixture of tar and pitch. It is then passed through a closing machine, ranged on a horizontal shaft, where eighteen strands of charcoal wire are laid on. Then it passes through a second mixture of tar and pitch, for the purpose of being protected from rust while waiting to be shipped ...[30]

Fleeming Jenkin, at the age of twenty-four, was much more than a 'philosophic assistant', for he had become Newall's chief engineer and electrician, designing machinery for paying out and picking up cables, overseeing the construction of those machines, fitting up cable ships and in charge of electrical testing. His role combined production management and technical development, the whole condensed into a seemingly impossible time scale which tried to meet the deadline of a brief cable-laying season and satisfy the impatience of investors. As in any new industry founded on an emerging technology, competent practitioners were in short supply. Indeed, some of the bitter feuds within the industry reflect the fact that there was no clear agreement of what constituted competence. But the acute shortage of electrical engineers presented Jenkin with an opportunity to shine, and within two years, his renown was such that he was called as an expert witness by the Galton enquiry in 1859.[31] 'His gifts', says Stevenson, 'had found their avenue and goal.'[32]

In letters to Anne Austin, Jenkin describes the responsibility and autonomy which were given to him, and the joy he felt in his 'bloodless, painless combat with wood and iron'.[33] Yet ceaseless attention to every detail was required, for he met with constant frustrations which drove him 'nearly mad with vexation ... I look upon it as the natural order of things, that if I order a thing, it will not be done – if by accident it gets done, it will certainly be done wrong; the only remedy being to watch the performance at every stage'.[34] As always with Jenkin, he endured the exceptional demands and managed to balance them with compensations. He wrote to Anne: 'The whole of the paying out and lifting machinery must be designed and ordered in two or three days,

---

[30] *Gateshead Observer*, 2 May 1857.
[31] *Minutes of Proceedings of the Institution of Civil Engineers*, LXXXII (1885), p. 365. The Galton committee is further discussed below, p. 53.
[32] RLS, p. lix.
[33] RLS, p. lxxvi.
[34] RLS, p. lxxvi.

and I am half crazy with work. I like it though: it's like a good ball, the excitement carries you through'. He complained that he often stayed at the works until ten or eleven at night, though he also found pleasures there: 'I have a nice office to sit in, with a fire to myself, and bright brass scientific instruments all round me, and books to read, and experiments to make, and enjoy myself amazingly. I find the study of electricity so entertaining that I am apt to neglect my other work.' Returning to the theme, he described to her 'some charming electrical experiments. What shall I compare them to – a new song? a Greek play?'[35]

Jenkin met William Thomson through Lewis Gordon in 1858.[36] Gordon visited his former colleague at Glasgow University to inspect cable-testing and signalling apparatus which Thomson was making for use on the Atlantic project. Jenkin was summoned by telegraph from Birkenhead, and spent a week with Thomson in his classroom and laboratory. Each was impressed with the other, and there began a professional collaboration which lasted until the end of Jenkin's life. Thomson later wrote:

> I was much struck, not only with his brightness and ability, but with his resolution to understand everything spoken of, to see if possible thoroughly through every difficult question, and (no *if* about this!) to slur over nothing. I soon found that thoroughness of honesty was as strongly engrained in the scientific as in the moral side of his character.[37]

Jenkin's interest in broader scientific questions, specifically on dynamics and physics, impressed Thomson. He also noticed the way in which Jenkin had seized his opportunity to experiment at the Birkenhead factory.

> Thus he began definite scientific investigation of the copper resistance of the conductor, and the insulating resistance and specific inductive capacity of its gutta-percha coating, in the factory, in various stages of manufacture; and he was the very first to introduce systematically into practice the grand system of absolute measurement founded in Germany by Gauss and Weber.[38]

The collaboration with Thomson centred upon the testing of conductivity and insulation of submarine cables, and the speed of signalling through them. Developing Faraday's discovery of the existence of specific inductive capacity, Jenkin was the first to measure correctly the specific

---

[35] RLS, p. lxi.
[36] The date of their first meeting is given as 1859 by Thomson, but the first surviving letter from Jenkin to Thomson dated 12 January 1859 suggests that they were already well-known to each other: Kelvin, Glasgow, J2.
[37] Colvin and Ewing (eds), 1887, I, p. clv.
[38] Colvin and Ewing (eds), 1887, I, p. clvi.

inductive capacity of any substance, through observation of submarine cables at the Birkenhead factory.[39]

Jenkin's first telegraph voyage was on the *Elba* in 1858, to retrieve Brett's two failed cables between Sardinia and Algeria.[40] The young engineer had designed, built and tested the picking up gear, consisting of a large pulley at the bow, a drum around which the cable passed six times before being deposited in the hold, and a small steam engine to drive the drum.[41] He had also brought some new paying-out equipment to experiment in shallow water. His duties were to maintain and adjust the machinery, and his side of it went well.

> Sixty, seventy, eighty, a hundred, a hundred and twenty revolutions at last, my little engine tears away ... I am very glad I am here, for my machines are my own children, and I look on their little failings with a parent's eye and lead them into the path of duty with gentleness and firmness. I am naturally in good spirits, but keep very quiet, for misfortunes may arise at any instant ...[42]

Liddell was on board, along with Louis Loeffler[43] and Frederick Charles Webb, wryly described by Jenkin as 'the two men learned in electricity'.[44] Loeffler worked for Siemens and Halske, engaged by Newall as electrical and consulting engineers to carry out electrical testing and assist with cable-laying on a number of expeditions during 1858 and 1859.[45] The tone of Jenkin's letters to Anne suggests that he was frustrated by Loeffler, who had difficulty locating faults on the line and issued conflicting instructions to Newall's men.[46] Loeffler and Webb argued about where the faults were; Liddell, Webb and the captain gave contradictory orders to the crew. Failure to agree technological solutions was compounded by a lack of established protocol among the engineers and electricians, the contractors and direct employees – who was actually in charge, and who was answerable to whom? There were also commercial tensions and attempts at cost-cutting which ultimately proved to be

----

[39] W. Thomson, 'Obituary Notices of Fellows Deceased', *Proceedings of the Royal Society*, XXXIX (1885), pp. i–iii.

[40] RLS, p. lxxv.

[41] RLS, p. lxxv.

[42] RLS, p. lxxxiii–lxxxiv.

[43] Johann Carl Ludwig Loeffler, whose name was anglicised to John Charles Lewis Loeffler when he sought admission to the Institution of Civil Engineers in 1865, was one of eight founders of the Society of Telegraph Engineers. He left £1.5 million in 1907: Appleyard 1939, pp. 29–30.

[44] RLS, p. lxxxiv. Stevenson disguised the names of Loeffler and Webb.

[45] Pole 1888, p. 117. Loeffler had been employed by the Siemens in Berlin, and later became managing director of Siemens Brothers in Britain: Pole, p. 163. Jenkin was one of his supporters when he joined the Institution of Civil Engineers in 1865.

[46] RLS, pp. lxxxvi.

false economies. Jenkin's machinery was adequate to pick up the smaller cable, but could not handle a larger one: 'The gear employed to take [the big cable] off the drum is not strong enough ... Luckily for my own conscience, the gear I had wanted was negatived by Mr Newall. Mr Liddell does not exactly blame me, but he says we might have had a silver pulley cheaper than the cost of this delay.'[47] He quickly improvised with part of the paying-out machinery and within hours had solved the problem. 'You would think someone would praise me; no – no more praise than blame before; perhaps now they think better of me, though.'[48]

Webb, although only five years older than Jenkin, had already a long experience of telegraph engineering. He had been soundly educated in, amongst other subjects, mathematics and mechanical drawing, and first qualified as a marine surveyor. Leaving the Royal Navy in 1845, he had worked as a civil engineer, mainly in railways, but soon joined the Electric Telegraph Company. From about 1850 until his retirement in the late 1870s, he concentrated upon telegraphy, mainly submarine, working for many of the leading companies and involved in most of the major cable-laying projects of the time.[49] Webb was employed for a year by Newall as assistant engineer, which took him on the *Elba*'s voyage in 1858.[50] Later he tested the failed 1858 Atlantic cable, showing that the distance of a fault could be roughly estimated by the change in resistance when the current was rapidly reversed. He was involved in experiments on behalf of the Galton enquiry, and gave lengthy evidence to that committee in 1860, as well as demonstrating his 'coil current', sometimes known as Webb's current, to Faraday. He wrote a number of learned papers – 'On the Practical Operations Connected with Paying out and Repairing Submarine Cables', delivered to the Institution of Civil Engineers in February 1858; several published in the *Philosophical Magazine*; and a collection from *The Electrician* during 1861 to 1865, which were republished in a volume entitled *A Treatise on Electrical Accumulation and Conduction*.[51] Jenkin credited Webb with the suggestion that the position of a fault in a cable could be detected by measuring the charge and resistance, an idea which had stimulated some of Jenkin's own experiments.[52] As to Webb's observations about differing behaviour in coiled and straight cables, Jenkin was more tactful in his public evidence to the Galton Committee than he had been privately to Thomson:

---

[47]   RLS, p. lxxxviii.
[48]   RLS, p. lxxxix.
[49]   *Electricians' Directory*, 1886, p. 156;   *The Electrician*, 1900, p. 84. Webb committed suicide in 1899.
[50]   See Webb's application for membership of the Institution of Civil Engineers, 1868.
[51]   *Electricians' Directory*, 1886, pp. 157–58.
[52]   Galton committee, p. 141.

Webb mentions the return currents in the *Engineer* of Saturday before last – he has done very little – gives I think a mistaken account of them says they only occur when a cable is coiled up and details a fallacious experiment on the subject with the Atlantic induction coils ...[53]

Differences with Webb were evident when they worked together on the 1858 expedition. Jenkin was 'uncommonly idle' and bored, pushed aside, for 'the buoying and dredging are managed entirely by W[ebb], who has had much experience in this sort of thing; so I have not enough to do, and get very homesick.'[54] Yet in spite of these disharmonious relationships, the disagreements about technical solutions, the problems with equipment, the delays and the unclear chain of command, by early July the job was completed and considered a success.

Our work is done: the whole of the six-wire cable has been recovered; only a small part of the three-wire, but that wire was bad and, owing to its twisted state, the value small. We may therefore be said to have been very successful.[55]

Newall's next project was the Dardanelles to Alexandria cable for the Levant Telegraph Company.[56] Jenkin was not present at the first attempt on this line, which reached its final stage, from Candia (now Iraklion, Crete), and was within seventy miles of Alexandria in November 1858 when the insulation failed. The cable had to be cut and it was too late in the season to try again.[57] Despite some uncertainty over whether the failure to meet a deadline of 24 December 1858 meant that Newall had forfeited his exclusive right to the route between Egypt and the Ottoman Empire, and had also lost a linked concession to lay a cable down the Red Sea, another attempt was to be made to reach Alexandria the following spring.

During 1858, William (C.W.) Siemens had established a factory in London for the express purpose of experimentation. His older brother Werner, head of the family firm, already owned a works in Berlin, and had himself made a distinguished contribution to submarine telegraphy through a pioneering study of gutta percha.[58] William Siemens wanted his own facilities in London, where he had lived since 1852, and opened his factory at Millbank

[53] Kelvin, Glasgow, J15, Jenkin to Thomson, 1 September 1859.
[54] RLS, p. xcii.
[55] RLS, p. xcv.
[56] The line was to link the Dardanelles with the Greek island of Khios, from where lines went to Izmir, on the Turkish mainland, and to Siros, an island south of Athens which was already connected to Crete. The final stages were between Athens and Siros, and Crete and Alexandria: National Maritime Museum, TCM/7/2.
[57] Newall 1860, p. 10.
[58] G. Tweedale, 'Sir Charles William Siemens', in Jeremy (ed.) 1986, V, p. 158.

Row, near Lambeth Bridge, for this purpose. There he employed between eighty and a hundred workmen on the manufacture of telegraph instruments and small components such as batteries and insulators. After moving to Woolwich in 1863 the Siemens company diversified, producing dynamos, pyrometers, galvanometers, and railway telegraph equipment, so that by 1891 it accounted for about a third of total British electrical and telegraphic production.[59] The Millbank factory had served its purpose for William Siemens, providing him with privacy and close control to carry out wide-ranging experimental work.[60]

> In these premises were elaborated the details of many electrical novelties, of furnace arrangements, of the refrigerating machines, the pyrometer, the resistance measurer, the bathometer, and of many other novelties, which went successfully before the world; and here were strangled in their birth any unfortunate offspring of Mr Siemens's brain which had not the power of competing in the struggle for existence.[61]

Jenkin's work was moving in a similar direction, as he became more closely involved with Thomson in developing telegraph instruments. Thomson was interested in all aspects of telegraphy, including cable manufacture and laying, but had also set himself the task of improving sending mechanisms and techniques, and instruments to receive and record messages. In short, he was concerned with developing the means to send and receive intelligible signals at an acceptably rapid rate.[62]

Thomson focused upon a problem which became known as 'retardation of the signals'.[63] The difficulty lay not in the speed of electrical currents in the cables, which was known to be very rapid indeed, but with the rate at which letters and words could be transmitted. This retardation became evident on the earliest attempts at commercial submarine telegraph lines. During laying, the first cross-Channel cable between Dover and Grisnez in 1850 was tested only for electrical continuity.[64] When the time came for signals to be received from Dover, it was assumed that something was amiss with the operator, for

> *letters* came, but they were so mixed that it was in many cases impossible to make any sense out of them ... The more the operator tried to control the letters the more erratic they became. At last it was suggested that the success attending the laying of the wire had caused the

---

[59] Tweedale in Jeremy (ed.) 1986, V, p. 159.
[60] Pole 1888, pp. 118–20.
[61] Pole 1888, p. 120.
[62] Sloan 1996, p. 72.
[63] See the account of Willoughby Smith, 1974 [1891].
[64] This cable was never brought into commercial use.

champagne to circulate so freely that the persons in the shore station at
Dover did not know what they were doing.[65]

Later it was confirmed that messages were being correctly sent, yet only
unintelligible, chaotic signals could be received. The immediate solution, to
restrict the rate at which operators worked, limited traffic so much that the
line's commercial viability was threatened. Willoughby Smith lamented many
years later that the phenomenon of electrical induction had not been understood
sooner. Looking back at the earliest undersea cables, Smith identified failures
in testing and quality control, as well as in scientific understanding, as
instrumental to their poor performance. All these were fields in which Jenkin
and Thomson, and their later partner Cromwell Fleetwood Varley, would
become pre-eminent.

By the 1850s, when retarded signals were causing such problems for
submarine telegraphy, the overland branch of the industry had developed, in
less than twenty years, into a sophisticated technological system. Signals could
be transmitted more or less automatically and at relatively high speeds, and
with the advent of printing receivers it was no longer necessary to read Morse
code. What was happening in the conductors, and, where used, in the insulating
envelopes, was a matter of indifference. If problems were experienced in
transmitting intelligible signals, usually electrical relays could be inserted in the
lines, and telegraph lines thus extended as far as necessary.[66] For most
electricians of the time,[67] the problem of signalling came down to 'strength of
the electricity', in other words, current. It was believed that if enough current
were transmitted through the wires, there could be no difficulty in operating a
receiver. But with the advent of the new and complex submarine branch of
telegraphy, it soon became plain that the problem of retardation was much
more profound than previously understood, and that stronger currents would
not solve it.

Michael Faraday was the first to explain the retardation phenomenon.
The problem had again arisen on the first working cross-channel line, one of
twenty-two miles laid between Dover and Calais in 1851, and then in three
cables, each 100 miles in length, connecting Orford Ness with Holland during

---

[65] Smith 1974 [1891], p. 9.

[66] As early as 1836–37 Samuel Morse had recognised the potential of relays.
According to George Prescott in *Electricity and the Electric Telegraph* (New York, 1877), pp.
459–60, Morse had said, 'Suppose that in experimenting on twenty miles of wire we should
find that the power of magnetism is so feeble that it will but move a lever with certainty a hair's
breadth;   that would be insufficient, it may be, to write or print, yet it would be sufficient to
close and break another or a second circuit twenty miles further, and this second circuit could
be made in the same manner to break and close a third circuit twenty miles further on, and so
round the globe.'

[67] The word 'electrician' was commonly used to describe those now classed as
electrical engineers.

1853. Faraday realised that a submarine cable, which was formed from a central copper conductor surrounded by an insulating envelope of gutta-percha, armoured on the outside with iron wire rope, formed an electrical capacitor.[68] When a pulse of electricity generated by a key was sent into a telegraph cable, there were – in simplified terms – two processes at work: an electrical current through the core, and an accumulation of charge in the capacitor. The net effect was that it took some time before the effects of a pulse applied to the input end of the cable became evident at the output. If successive pulses were too close together, confusion would result. William Thomson's theoretical study of transmission through insulated cables concluded that a pulse which started out more or less rectangular in form would emerge rounded and elongated, Figure 3.2. Elaborating his ideas for the Galton enquiry, Thomson showed how a series of pulses would appear at the output of a long submarine cable, Figure 3.3.[69]

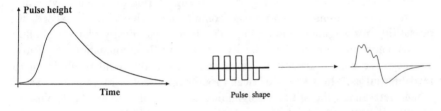

Figure 3.2: Pulse   Figure 3.3: Input and Output Pulses

Thomson had demonstrated that three parameters were involved in signalling speeds: the length of the cables; the conductivity of the copper core; and the specific inductive capacity of the insulating envelope. While the properties of the copper and the insulator were important, the length of cables exercised the strongest control on signalling speeds, for Thomson found that retardation of the signals was proportional to the square of the cable's length – his 'Law of the Squares'. Speed could be recovered by increasing the size of the conducting cores, but this would reflect back upon the design and construction of the system. The consequential increases in size and weight of cables would require bigger ships, heavier paying-out machinery, larger cable

---

[68] A capacitor is a device that can store electrical charge, the first practical example of which was the Leyden jar.

[69] This figure is adapted from those given between pages 134 and 135 of the Galton report. For a detailed treatment of the theory of signal transmission along submarine cables see Bright 1974 [1898], pp. 525–80, and for a modern interpretation Smith and Wise 1989, pp. 446–58.

making machinery and so on, throwing into question the economic viability of transoceanic telegraphy. The only other possible means of increasing signalling speeds through improving properties of the cable centred upon reducing impurities in the copper core to increase conductivity, but gains there were very limited.[70] Little could be done to improve the specific inductive capacity of the insulator.[71]

Thomson's hypothesis did not meet with universal approval. Wildman Whitehouse, electrician to the Atlantic Telegraph Company, would not accept the 'Law of the Squares', and produced experimental results in support of this view at a British Association meeting in 1856.[72] While Thomson, along with George Gabriel Stokes, kept faith with the theory, there were problems in verifying the law precisely: although the timing of input signals was straightforward, it was much more difficult to define and determine the timing of exit signals.[73] Whitehouse's measurements turned out to be flawed in this respect, and it was Fleeming Jenkin himself who eventually produced an unequivocal proof of Thomson's 'Law of the Squares'.[74]

In approaching a problem which was both scientific and practical, Thomson and Jenkin, by welding theory to engineering experience, were able to produce solutions of enormous commercial value. They realised the possibility of designing sending keys to attain economically viable signalling speeds Thomson showed theoretically that a suitable combination of positive and negative pulses, accurately timed and conjoined, would result in sharp, readable pulses.[75] In a series of neat experiments, Jenkin investigated factors which determined the speed of signalling and was able to verify Thomson's theory. He confirmed that several words a minute could be sent through cables thousands of miles in length.[76]

These results were not fully available in time to be of use on the first functioning transatlantic cable in 1858. This line worked for only a month, but long enough to emphasise the very slow rate of transmission then possible. By one account, Queen Victoria's congratulatory message to President Buchanan,

---

[70] Work in this direction did result in improved refining techniques, and in better quality control systems.

[71] Alternatives to gutta percha were generally found to be inadequate. India-rubber, for example, had better electrical properties but was unproven under the conditions experienced by deep sea cables. Jenkin could not recommend its use.

[72] Sloan 1996, p. 70.

[73] See for example the discussion in W. Thomson, 'On the Theory of the Electric Telegraph', *Proc. Roy. Soc.* VII (1855), pp. 382–99, especially pp. 397–9.

[74] This was contained in Jenkin's 1859 paper, 'On the retardation of signals through long submarine cables'.

[75] Hempstead 1998, pp. 40–41.

[76] 'Words a minute' had to be defined. Roughly, a word had a length of about eight dots and dashes. Five words a minute would therefore average forty pulses, in modern terms rather less than one bit per second.

ninety-eight words, took sixteen hours to send, and the President's 149-word reply a further ten hours.[77] The inherent electrical properties of the cable were not solely to blame for this, for the overall quality of the line was poor and the instruments inadequate. By the end of the decade, as the Galton committee began its deliberations, Thomson and Jenkin – and separately, Varley – were beginning to produce sending keys embodying the new theoretical and empirical knowledge, and were also applying themselves to the design of an automatic sending device. Over time, the coupling of automatic senders with Varley's 'curbing capacitors', and the introduction of duplexing, or two-way traffic on the line, effectively solved the problem of 'retardation of the signals'. Speeds on the transatlantic cable increased from about five words a minute in 1866, to thirty words a minute without duplexing by 1890.[78]

Soon after their first meeting, Jenkin was actively engaged upon testing Thomson's instruments in Newall's factory and at sea. He planned a thorough trial of Thomson's marine galvanometer during cable-laying in the summer of 1859. Thomson's initial breakthrough in telegraph instruments had been the mirror galvanometer, a receiver of extreme sensitivity which used a beam of light reflected off a mirror to enhance weak incoming signals. It was patented in 1858.[79] Using this, a 'mirror clerk' could read up to twenty-five words a minute, dictating the message to a second clerk.[80] The mirror galvanometer's very sensitivity, though, made it unsuitable for use on board ship during laying, when galvanometers were used for electrical testing and as receivers to try the function of the cable as it was laid. One problem was the ship's motion. The instrument also suffered the same magnetic problems which interfered with the working of ships' compasses – the changing magnetic field of the earth, and the influence of the ship's iron hull. Thomson developed a marine galvanometer which would work under these conditions, by suspending the mechanism to isolate it from the movement of the ship, using magnets to compensate for external influences, and in one version encasing the galvanometer in an iron shield.[81] Jenkin may have coined the term marine galvanometer, and certainly suggested modifications and improvements to Thomson's basic instrument:

---

[77] G. Saward, *The Trans-Atlantic Submarine Telegraph* (London: privately published, 1878), p. 34. Saward's account is contradicted by the journalist W.H. Russell who claims sixty-seven minutes for the Queen's message: Russell 1972 [1865], p. 26.

[78] These are very approximate figures. From 1858 to 1873, speeds could at times reach twenty-five words a minute, though the average was between seven and thirteen. The heavy cable of 1894 averaged fifty, with peaks of up to ninety words a minute: Headrick 1991, p. 32.

[79] Sloan 1996, pp. 77–79.

[80] Headrick 1991, p. 32.

[81] Sloan 1996, p. 77.

[They] should be what I call marine galvanometers – that is to say the
magnet should be suspended top and bottom and directed by a steel
magnet adjustable by screw – with frame lamp and scale attached ...[82]

Thomson's mirror galvanometer served the submarine telegraph industry for
fifteen years until superseded in 1873 by his siphon recorder.[83]

Jenkin experimented with a land galvanometer while awaiting delivery of
the marine instrument by steamer from White and Barr of Glasgow, the
makers, in March 1859.[84] Once delivered, the marine galvanometer, Jenkin
reported back to Thomson, 'appears to work beautifully'.[85] But on expedition,
he knew that he would not enjoy the same autonomy to experiment as in the
factory. He told Thomson: '... I am not sure how far it can be used during
submergence for I believe Messrs Siemens will have the testing at this time in
their hands and I shall not be allowed to interfere.'[86] The division of
responsibilities between Newall's contractors, Siemens, and the company's
direct employees remained uncertain. Jenkin had tried without success to
clarify his position, and wrote to Thomson:

> I quite see the necessity of some such arrangement as you describe to
> ensure the immediate discovery of a fault. At first I thought I was to have
> some voice in the testing and about three weeks since endeavoured to
> arrange such a plan with Messrs Siemens – They did not think I ought to
> interfere and my electrical position remains extremely undefined.[87]

After he had seen off the *Elba* – he was to meet the ship at Syra (Siros) in
late April – Jenkin stayed at Birkenhead, continuing experiments to locate the
position of faults in cables. He wrote to Thomson that he had been astonished
by the results 'for it appeared that the charge was nearly the same for the same
total resistance whether that resistance was made up of gutta percha covered
wire or of a common resistance coil'.[88] His efforts to experiment on board the
cable-laying ship were less successful. Upon his return three months later,
Jenkin apologised to Thomson that he had not been able to prepare a full report
on the galvanometer, nor could he offer any interesting information about the
electrical condition of cables before and after their submersion. Part of the
problem had been the lack of co-operation by Siemens:

[82]   Kelvin, Glasgow, J2, Jenkin to Thomson, 12 January 1859.
[83]   For technical details of the galvanometer see short discussion in next chapter.
[84]   Kelvin, Cambridge, J23, Jenkin to Thomson, 31 March 1859. The close and
longstanding relationship between Thomson and James White is examined in Clarke et al.,
1989, pp. 252–75.
[85]   Kelvin, Glasgow, J7, Jenkin to Thomson, 4 April 1859.
[86]   Kelvin, Glasgow, J6, Jenkin to Thomson, 22 March 1859.
[87]   Kelvin, Glasgow, J8, Jenkin to Thomson, 8 April 1859.
[88]   Kelvin, Glasgow, J9, Jenkin to Thomson, 19 April 1859.

I had taken the resistance and insulation of the Alexandria and Candia cable just before starting by your instrument, but our unfortunate loss of that cable prevented my getting any other information as during the paying out the cable was exclusively in the hands of Messrs Siemens assistants – in fact I had no room for my instruments, no batteries of my own and very very little assistance from Messrs Siemens men ...[89]

A further problem was that Jenkin had not been able to use the intended source of light for the galvanometer after the only crock of paraffin was broken on the voyage out. Although he had been able to set up a gas light instead, 'it was neither so convenient to manage nor so bright as the oil light would have been'.[90]

Once the cable-laying was underway, a series of problems arose. The shore-end in Alexandria was scarcely laid when the ship ran aground.[91] Then, after two thirds the length had been laid, the cable broke; Jenkin confided to Thomson that the cause was human error, that the man on the brake had not responded quickly enough when strain was placed on the cable, 'but probably Messrs R.S. Newall and Co do not wish this known'.[92] Jenkin was upset by the failure, but saw it as an opportunity to learn. The breakage had occurred during Liddell's watch and was thus not Jenkin's responsibility, but he told Anne: 'Though personally it may not really concern me, the accident weighs like a personal misfortune. Still, I am glad I was present: a failure is probably more instructive than a success ...'[93] He was to have more instruction, for the next task was to retrieve the Canea (Khania, Crete) cable, which broke constantly.[94]

As usual when faced with exacting demands, Jenkin derived some satisfaction from his surroundings, in this case in the beauties of the Mediterranean. He admired spectacular landscapes, reported back to Anne on ancient remains, and managed to sail and walk in the islands. Climbing through mountains near Candia, Liddell and Jenkin 'found a sure bay for the cable, had a tremendous lively scramble back to the boat. These are the bits of our life which I enjoy, which have some poetry, some grandeur in them'.[95]

The Crete to Alexandria cable had had to be cut and abandoned. Lionel Gisborne, from whom Newall had bought the concession to lay the Dardanelles-Alexandria line, later criticised Newall for failures in the manufacture of this cable. Newall countered that the cable had been 'subjected to the most searching tests' in his works, to a pressure of 1000 lbs. per square

[89] Kelvin, Glasgow, J11, Jenkin to Thomson, 21 July 1859.
[90] Kelvin, Glasgow, J11, Jenkin to Thomson, 21 July 1859.
[91] RLS, p. xcvii.
[92] RLS, p. xcviii;   Kelvin, Glasgow, J12, Jenkin to Thomson, 30 July 1859.
[93] RLS, p. xcviii.
[94] RLS, p. xcix.
[95] RLS, p. xcvii.

inch, with faults in the gutta percha emerging only after laying in 1,800 fathoms, where the pressure was 5400 lbs per square inch.[96] Newall in his turn blamed the Gutta Percha Company for inadequate testing of the core.[97] A large loss had been sustained by Newall and Co., ultimately amounting to 800 miles of cable. When Newall gave evidence to the Galton committee in August 1860 he was undecided about a further attempt upon the line, as the Turkish government had cancelled his monopoly the previous year.[98]

Jenkin was not on the ill-fated Red Sea voyage of 1859. The Red Sea telegraph was intended to provide a secure route from Britain to India, avoiding the practical and political problems of trying to run an overland telegraph across the Ottoman Empire. The submarine line, in six parts totalling 3500 miles, was to connect Egypt with the west coast of India. It was completed in February 1860, but the whole line, from Suez to Karachi, worked for only a few hours. No telegram ever travelled the whole distance, and it soon became apparent that there were serious problems with all six sections. The line had not been tested underwater before laying; the iron wires protecting it were too light and soon rusted, and worms ate through the insulation; and it had been laid taut on an uneven sea bed, so that when it became encrusted with barnacles the extra weight caused it to break.[99] Newall, Gordon, Gisborne, Werner Siemens and Webb were involved in the laying; William Siemens did not lay cables, on this or any other voyage, but was responsible for electrical testing. Consequently he avoided the wreck of the P and O steamship *Alma* on which the other engineers were returning to Britain. The ship went down on a coral reef in the Red Sea in June 1859. As a consequence, William Siemens' wedding to Lewis Gordon's sister, Anne, had to be delayed until Gordon and Werner Siemens could be rescued from the reef.[100] The families were already connected through Gordon's marriage to a distant cousin of the Siemens brothers. In business terms, though, Siemens and the Newall partners went separate ways after 1859, possibly as a result of Gordon's enforced retirement after the shipwreck.

Returning to Birkenhead after his own frustrating expedition of 1859, Jenkin continued to test Thomson's instrument:

> I have been extremely busy all this week, day and night almost, making experiments on your No 3 marine galvanometer which answers admirably. By means of a metronome I have got dots, lines, alternate lines and dots ... after transmission through various lengths of cable ...

[96]   Galton committee, p. 256.
[97]   Galton committee, p. 258.
[98]   Galton committee, p. 256. See also Newall 1860.
[99]   National Maritime Museum, TCM/7/2;   Pole 1888, pp. 168–69. See also Headrick, 1991, pp. 19–20.
[100]   Pole 1888, pp. 118, 122–23.

The spot of light will give 150 dots per minute without the least vibration – indeed when the batteries are clean I have had no difficulty on this score. There is no vibration on the arrival of current which has passed 370 miles – there is strong vibration on the arrival of current after 50 miles – I have not yet determined how short a length can be telegraphed through.[101]

Jenkin also tried out one of W.T. Henley's instruments, though apparently unimpressed by it, and he continued to work on improvements to signalling speeds.[102] At this time he began to consider protecting his and Thomson's work by patent: 'If you have not already patented these things and think I owe my discoveries if such they be partly to you would you patent them along with me?'[103] Soon afterwards he had developed a mechanical sender[104] and then an induction key, of which he wanted the instrument makers Barr and White 'to make a model rough and cheap in everything but essentials'.[105] He was not satisfied with another instrument maker, Elliott Brothers of the Strand, London:

The second galvanometer is not ready. Elliots [sic] have continued to make the most wonderful blunders with a pattern before their eyes and written instructions. They actually fitted lamps that would not burn paraffin![106]

Jenkin began to turn his attention to the economics of telegraphy as he was negotiating on behalf of Thomson with the board of the Red Sea company. He hoped to show the company that large savings could be achieved by closing a number of telegraph stations if they used Thomson and Jenkin's instruments. Jenkin hoped to arrange trials of the new instruments, and persuade the company to pay perhaps £750 a year for each station closed.[107] There were a number of meetings with the directors – 'it is rather a formidable thing being

---

[101] Kelvin, Glasgow, J12, Jenkin to Thomson, 30 July 1859.

[102] Kelvin, Glasgow, J17, Jenkin to Thomson, 9 September 1859.

[103] Kelvin, Glasgow, J20, Jenkin to Thomson, 15 September 1859.

[104] Kelvin, Glasgow, J24, Jenkin to Thomson, 24 September 1859.

[105] Kelvin, Glasgow, J33, Jenkin to Thomson, 20 February 1860.

[106] Kelvin, Glasgow, J37, Jenkin to Thomson, 24 July 1860. See Gloria Clifton, 'An introduction to the History of Elliott Brothers up to 1900', *Bulletin of the Scientific Instrument Society*, XXXVI (1993), pp. 2–7. Elliott's customers included Wheatstone, Maxwell and Latimer Clark, and the company later made transmitters, receivers, sounders, galvanometers, relays and keys, as well as testing equipment for cable manufacturers. Willoughby Smith's son was manager of the company from c.1873: H.R. Bristow, 'Elliott, Instrument Makers of London: Products, Customers and Development in the nineteenth century', *Bulletin of the Scientific Instrument Society*, XXXVI (1993), pp. 8–11. For these references and other information about Elliott Brothers we are grateful to Alison Morrison-Low.

[107] Kelvin, Glasgow, J31, Jenkin to Thomson, 20 January 1860.

cross-examined by a dozen worldly wise old gentlemen but I think I have done no damage yet'[108] – but he did not make progress, and ultimately his scheme was defeated by the failure of the Red Sea cable.[109]

Jenkin's experiences between 1858 and 1860 underline the problems with a still immature technology. Submarine telegraphy remained at a developmental stage, needing careful experiment which was not always possible on board ship or in a commercial workshop. But the Mediterranean and Middle Eastern expeditions did prove crucial in advancing the technology during the hiatus in Atlantic telegraphy between 1858 and 1865. While attempts to cross the Atlantic attracted more publicity, and the series of failures there unnerved many investors, work on the Mediterranean and Middle Eastern projects proceeded steadily. Cable technology improved, as did laying and testing techniques, and even though there were major failures among the Mediterranean and Red Sea projects, and the general public remained wary of buying submarine telegraph shares, those close to events – both engineers and large investors – were reassured that in principle the technology was viable, and the schemes promised to be highly profitable.

The British government was the main loser in the Red Sea failure, and this costly experience, along with the gradual demise of the Malta–Cagliari–Corfu line between 1857 and 1861, and the Malta–Alexandria telegraph's frequent breakdowns, brought the end of direct government involvement in submarine cables.[110] While the advantages of a cable network connecting the farthest outposts of empire were clearly seen – the Indian Mutiny in 1857 had played a major part in convincing Lord Derby's administration to back the Red Sea project, for it had taken forty days for an emergency request for more troops to reach London from Lucknow[111] – when Palmerston returned to power in 1859, with Gladstone as Chancellor of the Exchequer, there was no further question of such direct public financing of submarine telegraph companies. Instead, the government agreed to establish an enquiry into the technical aspects of submarine cables, a course which in the event served the industry better than any number of financial guarantees could have done. The enquiry was suggested by a respected Board of Trade technical expert, Captain Douglas Galton, who was subsequently appointed chairman. It was known as the Joint Committee, under the auspices of both the Board of Trade and the Atlantic Telegraph Company, and its deliberations were later described as 'the most valuable collection of facts, warnings, and evidence ever compiled concerning submarine cables'.[112] Galton saw an opportunity for a rigorous appraisal of

[108] Kelvin, Glasgow, J31, Jenkin to Thomson 20 January 1860.
[109] This episode is covered in more detail in chapter 4, below.
[110] Cain 1970, pp. 63–64.
[111] Headrick 1991, p. 19.
[112] Obituary of Sir Douglas Galton (1822–99), *The Electrician*, 1900, p. 80.

existing cable technology and manufacture, and believed that if quality and
reliability were improved, underwriting by government would no longer be
required as private enterprise would take up the job of extending the network.[113]
The Joint Committee met several times between December 1859 and
September 1860, and in 1861 published a report with recommendations on the
making, laying and working of undersea cables.[114]

Galton interviewed most of the leading submarine telegraph engineers of
the time, collecting evidence about the abortive attempts on the Atlantic, and
about the more and less successful cable-laying experiences in the
Mediterranean and elsewhere. The committee also invited reports and
commissioned experiments on a scale beyond the means of most laboratory
scientists.[115] Fleeming Jenkin was among these witnesses. He appeared before
the committee on 22 December 1859, submitting papers and details of
experiments, and subsequently wrote to Galton from Turin in September 1860
with further information and ideas, and an assurance that deep sea cables could
work.[116] As ever, he was loquacious – 'I talked a good deal the other day at the
Committee. What stuff it looks when reported by a shorthand writer, at least
mine did', he wrote to Thomson.[117] Jenkin's oral evidence to the committee was
based on a series of experiments carried out on lengths of cable at Newall's
works during the summer and autumn of 1859, when he determined the effects
of temperature on the resistance of gutta-percha and measured its specific
resistance.[118]

The first set of these experiments, concerned with temperature effects,
demonstrated a linear relationship with resistance, and established that pure
gutta-percha was a better insulator than gutta percha mixed with other
compounds. Many different combinations were available; Jenkin used
Chatterton's formula, a mixture of Stockholm tar, resin and gutta-percha in the
ratio 1:1:3. He concluded that while pure gutta percha possessed superior
electrical properties, Chatterton's compound allowed better adhesion between
the insulating envelope and the copper cores. His determination of the specific
resistance of gutta-percha, the first made on that material, and the first ever
such measurement for a non-conductor, made ingenious use of existing
apparatus and provided quantitative knowledge to inform the design of cables

---

[113] Cain 1970, pp. 54–5.
[114] *Report of Board of Trade Committee to inquire into best Plan for Construction of
Submarine Telegraph Cables*: PP (HC) 1860 [2744] LXII.591.
[115] Hunt 1994, p. 54.
[116] Galton committee, p. 145.
[117] Kelvin, Glasgow, J26, Jenkin to Thomson, 5 January 1860.
[118] Specific resistance is the resistance of a defined quantity of material. For a
description of Jenkin's method see Appendix 14 of Galton's report, pp. 464–81.

for different purposes.[119] Almost as an appendix to this, Jenkin presented a method by which the specific resistance of gutta-percha could be compared with that of copper, covering a range of twenty orders of magnitude, a remarkable result.[120] Jenkin also presented to Galton his verifications of the theoretical conclusions published by Thomson in 1856. Two of these were particularly important: the proof of an inverse relationship between signalling speed and the square of the length of a cable, all things being equal; and the fact that increased input voltage had no effect on the rate at which messages could be transmitted. Wildman Whitehouse remained unconvinced of the truth of the 'Law of the Squares' and was still persuaded that higher applied voltages would 'force' signals along the Atlantic cables. Whitehouse's decision to apply hundreds, if not thousands, of volts to the 1858 cable had had catastrophic results, but he preferred to lay the blame upon the poor standard of the cable itself. Not until Jenkin's evidence to Galton in support of Thomson was there indisputable proof of the folly of Whitehouse's action.[121]

The letter to Galton from Turin covered the origin, determination and history of faults in submarine cables, and was sent in lieu of an intended paper on the same subject.[122] Jenkin's main concerns were quality control, and the types of faults manifest in existing cables. He suggested conditions which would ensure that a cable lasted, and noted the self-healing of some faults. His main conclusion, though, was that the then state of knowledge was insufficient for any certainty about eliminating cable failures. Lengthy observations on the pathology of faults were needed, he thought, before that aspect of cable laying could be fully controlled.[123] Yet his own researches, measurements, observations and conclusions submitted to Galton were themselves significant contributions to the eventual success of submarine telegraphy.

The Galton enquiry, as well as exposing differences of opinion on technical issues, also revealed personal disagreement and resentment between submarine telegraph engineers. Jenkin's testimony did not pass without criticism. Willoughby Smith, another witness, was later dismissive of much of the evidence put forward to Galton, and specifically criticised Jenkin for hastily

---

[119] For a discussion of the role of gutta-percha and a brief account of Jenkin's experiments, see Hunt 1998.

[120] Hunt 1998, p. 97.

[121] Galton report, p. 142. See Jenkin, 'Experimental Researches...' in *Philosophical Transactions of the Royal Society* for 1862, for a fuller examination of these problems. See Hunt 1996 for a discussion of Whitehouse's role in early submarine telegraphy. Hunt makes clear that Whitehouse was not alone in his views of how best to manage the 1858 cable, but became a 'fall guy', sacrificed to protect the careers and reputations of others: p. 167.

[122] Galton report, p. 145.

[123] Galton report, p. 148.

publishing the results of experiments without properly ascertaining the materials of which cables were made.[124]

In his final report, thirty pages long, Galton compiled a list of all the submarine lines laid before 1859, along with an account of cable manufacture and laying. The report concluded with recommendations for the construction and operation of future lines. In a way it was a dismal picture, for of the many miles laid down since 1850 a large proportion had failed. On the whole, though, it was deep sea cables which had been unsuccessful, two thirds of which (in terms of length) were owned by four companies. The list identified many common mechanical causes of failure, usually breakages, and suggested that early successes with the cables between Dover and Calais, and Orford Ness and The Hague and Ostend, actually concealed serious shortcomings in knowledge and expertise. Galton's conclusions, although not stated overtly, can be summarised as: first, a need for careful quality control in the construction of cables; second, electrical units and standards to be defined and realised, so that individuals could use generally agreed resistance standards and common units of current and voltage; third, a general adoption of 'best practices' validated by experience, those developed by engineers such as Bright, Webb, Latimer Clark, Smith and Jenkin; fourth, close attention to the management of laying operations; and fifth, a need for standard signalling and receiving methods.

These conclusions emerged from the evidence given by those engineers who would develop submarine cables into a viable and reliable system over the following decade, so that the report was effectively a legitimating tool for already known best practice. The dominant ideas which Galton accepted came from a small coterie of engineers, but because his enquiry was exhaustive and wide-ranging, certain contentious issues were settled unequivocally and Galton's report proved to be decisive in shaping the future progress of deep-sea cables.

Because Fleeming Jenkin had not been directly involved in the highly publicised attempts to span the Atlantic, and because he was still relatively junior in his profession, his career with Newall did not bring him much financial reward, or wider fame. Some of those who were achieving celebrity – the likes of Charles Bright, knighted at the age of twenty-six when chief electrician to the 1858 Atlantic cable, and Webb and Smith – were only slightly older than Jenkin, but had gained earlier experience and were given opportunities to accompany the more prestigious expeditions. But the work with Newall and Liddell made Jenkin's name within the profession, and, along with his interest in science generally – 'full of intelligent eagerness on many particular questions of dynamics and physics' – excited Thomson's admiration

---

[124] Smith 1974 [1891], pp. 92–93. Willoughby Smith (1828–91) was employed at the time of the enquiry by the Gutta Percha Company and later by the Telegraph Construction and Maintenance Company.

from their first encounter. In Thomson's view, Jenkin 'became early known as an electrical engineer of high standing.'[125] The invitation to Jenkin, when still only twenty-six, to give evidence to the Joint Committee seems to confirm that opinion.

From 1860, largely through his collaborative work with William Thomson, Jenkin began to achieve wider recognition for his achievements in submarine telegraphy. While Galton's committee had resolved some fundamental questions relating to undersea cables, the technology still presented many intellectual and practical challenges. Standard units of electrical measurement remained to be determined. It was also clear that, for long cables to be laid and managed satisfactorily, much improved sending, receiving and testing instruments must be developed. Jenkin's opportunity came when he was asked to join in partnership with Thomson, patenting and marketing innovations in telegraphy, and by the beginning of 1860 he had decided to leave Newall and Liddell.[126]

[125] W. Thomson, 'Obituary Notices of Fellows Deceased', *Proceedings of the Royal Society*, XXXIX (1885), p. i.
[126] Kelvin, Glasgow, J32, Jenkin to Thomson, 3 February 1860.

Appendix: Thomson's Mirror Galvanometer

The land and marine versions of Thomson's instrument were closely related; both could react to very small currents, had low mechanical inertia, and as they had little or no damping, could respond quickly to changing currents. They were designed to mitigate the effects which long insulated cables had on the propagation of electrical pulses. The galvanometers were moving magnet devices, a common form in the 1850s, but were more or less independent of the effects of extraneous magnetic fields. The marine galvanometer was designed

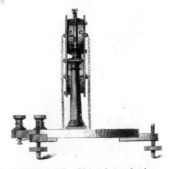

Figure 3.4: Front view of mirror galvanometer (Prescott, 1877, p. 151)

Figure 3.5: Side view of mirror galvanometer (Prescott, 1877, p. 152)

to function even in the most extreme conditions experienced by ships at sea.

There were many models of land mirror galvanometer: Figures 3.4 and 3.5 show two views of a typical working version. All were constructed upon the same basic pattern, with significant components and mode of operation as indicated in Figure 3.6. A current passing through coils C_1 and C_2 acted upon the small bar magnets, and the resulting deflection was detected by a spot of light reflected from the mirror. Each coil consisted of two windings of many turns, connected in parallel, making an extremely sensitive instrument. The magnets, made of high quality steel, were arranged so that the fields were opposed, so that the effects of any external fields were minimised while enough of the earth's magnetic field remained to act as a small restoring force. The

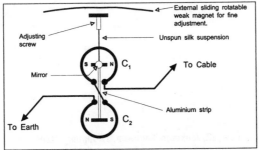

Figure 3.6: Astatic galvanometer, sketch

magnets were joined by a thin strip of aluminium and were suspended by unspun silk threads. A light beam, derived from an oil lamp, was directed at and reflected from a small mirror. Any movement on the mirror was displayed as a moving spot of light on a distant screen. The instrument, although sensitive, was easy to use and to adjust.

The marine galvanometer developed by Thomson was less sensitive than its terrestrial relation and sufficiently robust to be used on cable-laying expeditions. Figure 3.7 shows that Jenkin's ideas, suggested in a letter of January 1859, of a 'magnet suspended top and bottom, and directed by a steel

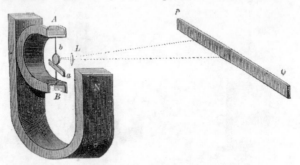

Figure 3.7: Internal arrangement of marine galvanometer
(Prescott, 1877, p. 154)

magnet adjustable by a screw – with frame lamp and scale attached', had been picked up by Thomson.[127] There was a single coil, in the centre of which were suspended a small lamp and mirror, which could be plane or a plano-convex lens silvered on the curved side – the latter produced sharper images on the distant scale. The suspension threads passed accurately through the centre of gravity of the magnet-mirror combination, so that the relative positions of coil, mirror and scale remained unchanged regardless of the instrument's motion. Apart from the scale, the whole instrument was enclosed in soft iron.

---

[127] Kelvin, Glasgow, J2, Jenkin to Thomson, 12 January 1859.

Chapter 4

An Electrical Engineer of High Standing[1]

The practice of electrical engineering in 1860 centred upon telegraphy. Submarine telegraphs, still in their infancy, presented the greatest technological challenges to engineers. There were already undersea lines in use which worked well – indeed, even the 1858 Atlantic cable, many times longer and deeper than any other of its time, had functioned briefly before breaking down completely. In general there was confidence among submarine telegraph engineers that the technical difficulties would be solved, although no consensus had emerged upon the best methods of insulating and protecting cables. It was also clear that laying and grappling techniques could improve, and that better working procedures and instruments – to transmit, receive and test lines – were needed. Contentious electrical issues had to be resolved, weaknesses within the system corrected. Until that happened, ventures into submarine telegraphy, especially attempts to lay long distance deep-sea cables, carried risks not only to the finances of investors, but also threatened the reputations of engineers.

Among electrical engineers, a small and close group working at a technological cutting edge, there was a perceptible distance between academic electricians and some of the more empirical engineers. To separate submarine telegraphy into two domains, the theoretical and the practical, would be to over-simplify, for a top-ranking scientist like Thomson could also be a respected practitioner, while Thomson and Jenkin's later partner, Cromwell Fleetwood Varley, starting from the engineering end of the spectrum, developed 'high scientific talent' to add to his practical competence.[2] But there were also prominent and influential telegraph engineers who mistrusted theoretical approaches to problem-solving, and had neither abstract understanding of electricity nor scientific rigour, although some did try to experiment.[3] The deliberations of Galton's committee heralded more

[1]   Sir William Thomson, 'Obituary of H.C. Fleeming Jenkin', *Proceedings of the Royal Society*, XXXIX (1885), p. i.
[2]   Smith and Wise 1989, pp. 677, 702. The quote is Thomson's.
[3]   See for example Jenkin's comments to Thomson about Webb: Kelvin, Glasgow, J15, 1 September 1859.

disciplined and methodical procedures, endorsing those who saw, as Thomson
had, that not only was theory an integral part of the development of telegraphy,
but also that telegraphy would inform scientific theory. There could be no
'double set of natural laws'.[4] To succeed with deep-sea cables, to achieve the
Atlantic crossing, to eliminate an unacceptably high risk from the business of
submarine telegraphy, required a symbiosis of scientific understanding with
best practice. (See Figure 4.1 for an illustration of the 'state of the art in 1866,
as visualised by the Illustrated London News's artist).

Figure 4.1: Testing the recovered cable of 1865 (*Illustrated London News*, 13
October 1866)

Fleeming Jenkin had chosen an inauspicious time to leave Newall and
Liddell, for the hopes of high returns from his association with Thomson
rested upon the agreement to license their instruments to the Red Sea and India
Telegraph Company. The final stage of the Red Sea line was completed in
February 1860, but it quickly became apparent that the cable was failing. By
March 1860 five of the six sections had broken down,[5] and despite attempts at
repair, four of these were completely defunct by April 1861.[6] However, some
of the cable functioned for eighteen months to two years, and, crucially for the
British government which had guaranteed the cable, each section tested well
for thirty days. The government was therefore bound by contract to pay the
Red Sea shareholders £36,000 per annum for the following fifty years – a 4.5

----

[4]   Smith and Wise 1989, p. 666.
[5]   Headrick 1991, p. 20; Cain 1970, p. 61.
[6]   Pole 1888, p. 169; National Maritime Museum, TCM/7/2.

per cent return on the capital of £800,000 – at an eventual cost to the exchequer of £1.8 million.[7]

For Jenkin there was no such financial cushion, and the uncertainty surrounding the Red Sea scheme during 1860 caused him great anxiety. He managed to gain admittance to see the Red Sea directors at the end of March, complaining of the continuing delay in reaching an agreement about the instruments – 'we had been kept waiting for a much longer time than we had been led to expect'. His report to Thomson about this meeting makes clear that all was far from well with the project, although Jenkin still expected the difficulties to be overcome. But the cost-saving potential of the instruments, the idea that money would be saved by closing telegraph relay stations, was being undermined by doubts about the line's reliability:

> One member said that they had reason to fear continual interruptions between the stations which must be temporarily supplied by steamer communication and that therefore it would be necessary to retain all the stations if not to form more – I answered that I did not anticipate much interruption – but even should the present stations be retained our system would permit double the speed to be maintained – they caught eagerly at this idea & said they imagined our chief advantage was to be obtained in long links only – I replied that we could with certainty promise them 12 words where they now got six ... they wished to know if we could get 20 ...[8]

The key patented by Jenkin and Thomson was designed to satisfy signalling protocols determined by Thomson and accepted by Galton. Jenkin's experiments had confirmed that by sending a correctly timed train of positive and negative electric pulses, the signal could be read much more easily and quickly, with obvious operating benefits. The key and its later derivative, the 1874 Thomson-Jenkin automatic sender, undoubtedly allowed signalling speeds to be increased. There were two problems, however, for its promoters in the early 1860s. In many cases the traffic on submarine lines was insufficient to warrant the application of mechanical or semi-automatic keys before the early 1870s, and even then only the North Atlantic cables tended towards saturation. The other difficulty was that the receiving instrument, the mirror galvanometer, although very sensitive was tedious to interpret and it is unlikely that it could have been useful at the sending speeds attainable with keys such as Jenkin and Thomson's.[9]

Jenkin had been hoping to try out his key on the Red Sea cable itself, and by demonstrating its effectiveness convince the directors that his financial

---

[7]   Headrick 1991, p. 20.
[8]   Kelvin, Glasgow, J35, Jenkin to Thomson, 30 March 1860.
[9]   For details of these instruments see Appendix Chapter 4.

argument was sound. He had also hoped that the board would pay his expenses
for the trip, perhaps by employing him for a time, although he was willing – 'if
the worst came to the worst' – to find the money himself. It seemed to Jenkin
that the problem lay in the directors' failure to see the potential of his
instrument – he could not try it until they agreed, and without such a trial they
would not appreciate its value.[10] In retrospect, it seems that the directors
delayed because they knew of the doubts surrounding the future of the line.
Another difficulty faced by Jenkin was that he did not wish to patent the key
before testing it, in case details were changed as a result of the trial. He wanted
to avoid the expense of taking out more than one patent, and wrote to
Thomson: 'I am therefore very anxious to hear of some line where we could
try it – can Latimer Clark not help us?'[11]

By July of 1860, Jenkin had found other work. To his great satisfaction
he was to lead an expedition for the first time, which presented the chance he
needed to test his and Thomson's latest instruments. On behalf of the French
government he was to repair the Bona-Spartivento cable between Algeria and
southern Italy. This was the line which Brett had failed to complete and which
Newall took on in 1857. Newall claimed that initially all four wires had
worked, though this was disputed by Brett and others. For most of the cable's
life only two wires functioned, and those two failed during 1860.[12] Despite this
history of problems on the line, Jenkin showed cautious optimism: 'I have
some hopes of success', he wrote to Thomson.[13] He stayed at Stowting in
September, *en route* to Paris and Turin, where he met the *Elba*.[14] By October
he was trying to repair the land line at Spartivento, where malaria was endemic
and it was considered unsafe to sleep on shore.[15] He revelled in his
responsibility, writing to his wife:

> Certainly being at the head of things is pleasanter than being subordinate. We
> all agree very well; and I have made the testing room into a kind of private
> room, where I can come and write to you undisturbed, surrounded by my dear,
> bright brass things which all of them remind me of our nights at Birkenhead.
> Then I can work here too, and try lots of experiments; you know how I like
> that! and now and then I read – Shakespeare principally ...[16]

[10] Kelvin, Glasgow, J34, Jenkin to Thomson, 28 February 1860.
[11] Kelvin, Glasgow, J34, Jenkin to Thomson, 28 February 1860.
[12] Galton committee, pp. 251–52.
[13] Kelvin, Glasgow, J37, Jenkin to Thomson, 24 July 1860.
[14] Kelvin, Glasgow, J42, Jenkin to Thomson, 18 September 1860.
[15] RLS, p.ci; Newall made the same point in his evidence to the Galton committee,
p.251.
[16] RLS, p. cii.

Though enjoying the command, Jenkin began to despair of achieving his objective:

> Decidedly I prefer being master to being man: boats at all hours, stewards flying for marmalade, captain inquiring when ship is to sail, clerks to copy my writing, the boat to steer when we go out – I have run her nose on several times; decidedly, I begin to feel quite a little king. Confound the cable, though! I shall never be able to repair it.[17]

Continuing his cable-fishing (as Stevenson put it), a week later he had found a break at the Algerian end of the line, which he was sure had been caused by coral fishers. Further out to sea, forty miles of cable were recovered, some of it from a depth of 1200 fathoms. The whole of the cable was covered in coral and marine animals, although these did not seem to have caused the failure. Jenkin sent specimens of the various creatures for identification to a Professor Allman, F.R.S., along with information about the depths at which they had been found. Fifteen varieties were identified, some from depths where it had been assumed life could not be sustained. A newly discovered *Caryophyllia*, a type of coral, was named *'Caryophillia borealis* Fleeming' after its finder.[18]

More fractures in the line were discovered, and repaired, but there was still no communication with Italy. Jenkin feared that the men left at Spartivento had fallen ill.[19] It later emerged that the problem lay with the cable, not the operators. On his return to England, six weeks later than anticipated, Jenkin wrote to Thomson from Stowting, apologising for his long silence and summarising his frustrating experiences in the Mediterranean:

> Meanwhile here is a catalogue of what I have *not* been able to do. I have not been able to repair the Bona and Sardinia cable – nor consequently to try the sending instruments. I have not been able to make any experiments on earth currents or to leave any instructions with Clerks about them as I have never come across a submarine cable in action.[20]

But the picture was not entirely bleak. Jenkin had been able to conduct 'a great many' experiments on the resistance of faulty cables, using the four conductors in the Bona cable, though the results were still unclear.

> I have made experiments on return currents, polarisation currents and fine currents without being able to find a trace of an earth current. The

---

[17] RLS, p. ciii.
[18] Wyville Thomson, *The Depths of the Sea* (London, 1874), pp. 26–28. Wyville Thomson had read Jenkin's journal of the voyage, since lost.
[19] RLS, p. cvi.
[20] Kelvin, Glasgow, J43, Jenkin to Thomson, 10 December 1860.

effects produced by these several sources of electricity are so complicated that I have been unable to draw any satisfactory conclusions from them.[21]

On other matters his opinions were more decided:

> The resistance test with the parallelogram on the contrary I find *perfect* – the marine galvanometers as galvanometers *perfect*. The resistance bobbins defective as I have always thought in arrangement; as you cannot follow changes of resistance quickly with them.[22]

He also told Thomson that he had 'prospect of good work with the Sardinian government inspecting and perhaps laying cables' – for it was clear by this time that the Red Sea scheme was in ruins, and Jenkin needed other employment.

Relief came in the form of a partnership with Henry Charles Forde (1827–97). Jenkin had met Forde while working for Newall on the Malta-Alexandria cable, upon which Forde represented the British government.[23] Forde's career had begun in railway and canal construction, and he was in partnership with Lionel Gisborne from 1854. Gisborne and Forde developed an interest in submarine telegraphs through their connection with the East India Company, and were appointed engineers to the Red Sea and India Telegraph Company in 1857 where they supervised the work of the contractor, Newall. Jenkin's partnership with Forde began in about January 1861, shortly before Gisborne's death, which occurred in March 1861.[24] Forde and Jenkin, described as telegraph and general engineers, operated from the same premises which Forde had shared with Gisborne in Duke Street, Adelphi – now called John Adam Street – off the Strand in London. It was in a district where many civil and telegraph engineers had offices, and 6 Duke Street was, like most of its neighbours, divided into small chambers which accommodated a number of architects and surveyors as well as several civil engineers.[25]

Forde and Jenkin's most noteworthy work was as engineers-in-chief for Reuter's, for the French Atlantic and the German Union Telegraph companies, and for the Rouen Waterworks; also they acted as consultants for the British Government when inland telegraphs were taken over by the Post Office in 1868.[26] The partnership ended in 1868 although their association continued less

---

[21]   Kelvin, Glasgow, J43, Jenkin to Thomson, 10 December 1860.
[22]   Kelvin, Glasgow, J43, Jenkin to Thomson, 10 December 1860. The 'parallelogram' referred to was a Wheatstone bridge.
[23]   Obituary in *Institution of Civil Engineers Proceedings*, pp. 365–6.
[24]   Smith 1974 [1891], p. 85.
[25]   *Electrical Trades Directory*, 1898, pp. 79–80; *Post Office London Directory*, 1863, trades directory – see 'Civil Engineers' list.
[26]   Obituary in *Institution of Civil Engineers Proceedings*, p. 366.

formally, and sometimes in conjunction with Latimer Clark, after Jenkin moved to Edinburgh. Forde subsequently went into business with Clark and later also with Herbert Taylor, overseeing the installation of 100,000 miles of submarine cable, much of it in the Far East and south Atlantic.[27]

Josiah Latimer Clark (1822–98) was an innovative practical engineer who also made a distinguished contribution to academic research. A report written by Clark in 1853 about underground telegraph cables – a technology closer to submarine than to overland telegraphy – was presented to the Galton Committee, of which he was a member. Clark was the first to notice the retardation of electric signals in submarine lines, and to demonstrate that currents of low tension travel as fast as those of high tension; he discovered how messages could be transmitted pneumatically, a method which was soon in widespread use; he patented a double cup insulator for overland telegraphs in 1856; he developed a material to protect cables, known as Clark's compound, in 1858.[28] In 1859, at a time when the transatlantic project was in abeyance, Clark replaced Sir Charles Bright as engineer to the Atlantic Telegraph Company, but resigned from this post in 1861 to become Bright's partner.[29] The pair carried out a series of highly regarded experiments in 1863 to test the effect of temperature upon the insulation of gutta percha cables.[30]

In February 1861, early in his association with Forde, Jenkin was in Paris, conducting a *post mortem* on parts of the Bona-Spartivento cable which had been retrieved. He wrote at length to Thomson detailing the faults.[31] All four wires had defects in the gutta percha which had evidently occurred during the production process, before they were stranded together. Because the tar covering had lasted well and provided a partial insulation, imperfections in the gutta percha had not been detected before the cable was laid. Another small fault turned out to be an air bubble filled with water, which could be located exactly only by touch. But the defects which Jenkin saw in Paris, while explaining the poor working of the cable, did not account for its ultimate failure. His experiments had been hindered by problems with the land lines at Spartivento, and then by inadequate marking of the four wires in various

---

[27] See the *Electricians' Directory* 1888, p. xvi; also 1893, p. xvii, and 1899, p. 78. Some records of this partnership are in the Science Museum Library, Muirhead collection. They became a public company, Latimer Clark, Muirhead and Co. Ltd., in the early 1880s: Kelvin, Cambridge, C92, Latimer Clark to Thomson, 22 December 1883.

[28] *Electricians' Directory*, 1885, pp. 94–5.

[29] New York Public Library, Wheeler collection, no.1934, autobiographical notes of Latimer Clark, 1875.

[30] The account of these experiments in the *Electricians' Directory* for 1885, p. 95, does not make clear where they took place, or whether Clark and Bright received financial support to conduct them. In 1853 Clark had experimented on the retardation of electric signals in submarine cables at the Gutta Percha Works in London, along with Michael Faraday: see Sloan 1996, pp. 62–4.

[31] Kelvin, Cambridge, J26, Jenkin to Thomson, 22 February 1861.

pieces presented to him for examination, making it impossible to match up the sections of cable.

In the absence of a full record, only an incomplete account can be given of Forde and Jenkin's work during the 1860s. One of their more important commissions was a report, published in October 1862, recommending a structure for the Persian Gulf cable to be laid and operated by the Indian government.[32] They applied lessons learned through the failures of Atlantic and Red Sea cables, and endorsing Galton's approach showed that solidly based knowledge, of the properties of materials and of electricity, allowed a submarine cable to be rationally designed. Examining in turn the core, the insulator and testing procedures, Forde and Jenkin approved the use of copper of high purity; they recommended a gutta-percha envelope even though india-rubber may have had a claim to certain superior properties; and they specified tests to be applied during manufacture, during laying and once the cable was in operation. Recognising commercial realities, the report made compromises between strength and electrical properties; learning from a decade's experience of submarine telegraphy, it concluded that as mechanical failures were by then rare, it was possible to design cables to take account of economic considerations such as the need for increased speed of signalling. Forde and Jenkin's advice embodied all Galton's 'best practices', and their report was adopted by the company, with successful results.

Also in 1862, Jenkin tested Hooper's india rubber core on behalf of Newall, finding the insulation 'excellent'.[33] Around that time, he was also giving advice on cables and potential signalling speeds to Cyrus Field, leading promoter of the Atlantic project, who was trying to revive that scheme. Jenkin had hopes that he and Forde, or he and Thomson, would be appointed engineers and electricians to the Atlantic Telegraph Company.[34] Forde and Jenkin declined an offer later in 1862 to serve as consulting electrical engineers to a proposed new company intended to refinance and run the works of W.T. Henley, the third of the major cable-making companies after Newall and Glass and Elliot.[35] Jenkin explained that a conflict of interests prevented

---

[32]   Institution of Electrical Engineers, SC Mss 73/2.

[33]   National Maritime Museum, Ms 88/078.

[34]   Kelvin, Glasgow, J45, Jenkin to Thomson, 13 March 1862.

[35]   This scheme, to establish The Telegraph Cable, Plant and Maintenance Company Limited, did not proceed. W.T. Henley, a successful cable-maker who was seriously under-capitalised, afterwards entered into a working agreement with The Telegraph Maintenance and Construction Company Limited (formerly Glass and Elliot), which lasted from 1864 until 1871. The company advanced working capital, passed on contracts, and bought cables from Henley. See the Science Museum Library, HEN 29, 'The Early Life of W.T. Henley, By Himself'.

their representing a cable manufacturer if they wanted to continue work for various telegraph companies.[36] During 1863 the partners were in discussion with Bright and Clark about working together on 'the Australian business', and there was also mention in letters about repairs to the Malta-Alexandria line, although it is unclear whether they carried out any work.[37] Then there is a break of three years in information about cable-laying and cable-repairing, suggesting that Jenkin's attention may have been directed at this time into his work with Thomson, improving instruments for the Atlantic crossing. The Forde-Jenkin partnership was not doing well at this time, and money worries were a constant anxiety for Jenkin. In 1865 he wrote from Duke Street to Thomson in response to a request for advice on a career in engineering from the son of Professor Forbes of St Andrews. Jenkin considered himself well-qualified to offer a view, having himself 'made practical trial of the difficulties a young engineer has to encounter'. Particular problems, he thought, faced aspiring engineers without an existing family interest to advance their career.

> Probably in saying this I am only saying what every member of any profession will say of his own who has himself had no 'interest' but in Civil Engineering there is this special disadvantage that a young man's ability is simply carried in most instances to the credit of his employer. The Great Engineers are really so many contractors for Engineering who do it at so much 'per mile' (for railways). They have a large staff indifferently paid, the members of which gain no credit outside a small circle of fellow workers. As a subordinate the pay is never high and it is excessively difficult to begin business on ones own account. There are very few small jobs and the few there are, are picked up by men who can take shares or have monied friends or have relations with the contractor etc. in fine through anything but merit.[38]

After the successful completion of the Atlantic telegraph in the summer of 1866, business improved. By September Forde and Jenkin had another large project in hand, acting for Reuters Telegraph Company during the laying of the cable from Lowestoft to Norderney in Prussia.[39] This cable, operated by the Submarine Telegraph Company on behalf of Reuter, formed part of the Indo-

---

[36] Kelvin, Glasgow, J47, Jenkin to Thomson, 12 December 1862.
[37] Kelvin, Glasgow, J48, Jenkin to Thomson, 23 July 1863; J50, Jenkin to Thomson, 23 September 1863.
[38] University of St Andrews Library, Forbes correspondence (in) 1865/62, Jenkin to Thomson, 5 July 1865.
[39] For an account of Reuter's involvement with land and submarine telegraphy, see Donald Read, *The Power of News: the History of Reuters, 1849–1989* (Oxford University Press, 1999).

European Telegraph Company's link from London to Karachi.[40] Jenkin, who was the only engineer on board representing Reuter, suffered a painful attack of rheumatism after catching a chill, as well as enduring the sea-sickness which afflicted him on every telegraph voyage.[41] During the same year, Jenkin was electrician on the voyage to lay Newall's cable from Khios to Crete, a distance of 180 nautical miles.[42]

In his evidence to an arbitration case in 1867, Jenkin said that he and Forde were engineers to the Crown Agents for the Colonies – 'it is a small business' – and also to Rogers of the Telegraph Company.[43] The following summer, as he was preparing to leave Forde for Edinburgh, he wrote that they expected to be employed 'to report to Government on the present condition of all the submarine cables in connection with their transfer'.[44] His comment refers to the transfer of inland telegraphs, including some underwater cables, into public ownership in 1868, and he did indeed test all those cables.[45]

Although Jenkin had missed a chance to distinguish himself alongside Thomson on the Atlantic expeditions of 1857 and 1858, and his career had suffered a setback through the failure of the Red Sea scheme in 1860, nonetheless submarine telegraphy still presented great opportunities to exercise his burning intellectual curiosity. According to Hunt, the failure of attempts during the 1850s to span the Atlantic raised 'questions that went to the heart of the future development of submarine telegraphy', and carried telegraph engineering into many new areas.[46] One of these was engineering specifications and standards, a field in which Jenkin was to make a lasting mark. In this as in many of his other activities from 1860, Jenkin worked closely with William Thomson. Thomson has been described as 'in part scientist, inventor, expert adviser and consultant, and even entrepreneur' but with his priorities focused upon carrying out a technology programme rather than a business plan.[47] Sloan sees Thomson as a prototype of the late twentieth-century 'scientific scholar', an academic concerned with translating his scientific research into industrial applications which are also business opportunities. In this analysis, the success achieved by the partners was attributable much more to their technological talents than to their business acumen.[48] Yet Thomson and Jenkin were obliged

---

[40]   Haigh 1968, p. 200.
[41]   RLS, pp. cvii–cviii; National Maritime Museum, TCM/7/2.
[42]   National Maritime Museum, TCM/7/2.
[43]   National Maritime Museum, ms 88/078. It is not clear which telegraph company is referred to.
[44]   Kelvin, Glasgow, J88, Jenkin to Thomson, 13 August 1868. The partners had offered their services for this work in 1867: PRO, T108/4, 67/767.
[45]   Obituary in *Institution of Civil Engineers Proceedings*, p. 366.
[46]   Hunt 1994, p. 52.
[47]   Sloan 1996, p. 18.
[48]   Sloan 1996, p. 19.

to devote a great deal of attention to business matters. To convince telegraph companies of the value of their various innovations, commercial as well as technological advantages had to be demonstrated. Furthermore it was necessary to pay close attention to patents and agreements with the telegraph companies to ensure that the partners received adequate compensation for their work and development costs. While companies felt that they had suffered great losses during the 1850s though the mistakes of some engineers, Thomson and others had shown faith in the future of submarine telegraphy by personally financing experiments, giving their services free, even on occasion covering their own travelling and accommodation expenses.[49] After 1860, as the technology of undersea telegraphs found a clearer direction, companies adopted more businesslike procedures. As an example, specifications in cable contracts became much more precise, designating materials and insisting upon certain electrical characteristics.[50] It was inconceivable that a dispute like that of 1857 regarding the wrongly twisted cable could recur. (Figures 4.2 and 4.3 indicate how, by 1865, considerable expertise and control were being exercised in the mechanical engineering and management of submarine telegraphy). In these changing circumstances, engineers had to devote close attention to business details. Jenkin's admiration for Thomson meant that he was willing to take on the burden of administering the partnership's affairs to give Thomson more time for technical work. The result of this was to boost the senior partner's career at the expense of the junior. Jenkin tended to defer to Thomson, while Thomson could be high-handed and patronising to Jenkin. Yet Jenkin's work was of central importance to the partnership.

Partnership terms were negotiated while the Red Sea scheme still promised large returns. Jenkin's proposal to Thomson early in January 1860 was that the first £3000 of clear profit from their patents be divided equally – 'I think the profits from the Red Sea scheme will in all probability be much more than three thousand pounds' – to reflect other opportunities lost to Jenkin by his engagement with Thomson. The following £3000 would be shared two thirds: one third in Thomson's favour, and thereafter profits would be evenly distributed.[51] When Cromwell Fleetwood Varley entered the partnership in 1865, the new agreement was to divide profits so that Thomson would take a half, Jenkin and Varley one quarter each. In the event, it was many years before profits from patents and consultancy services reached the dizzy heights anticipated before the financial calamity of the Red Sea project. The partners' achievements during most of the 1860s are better measured in technical rather than financial terms.

[49]   Kelvin, Glasgow, V20, Varley to Sir Richard Glass, 10 June 1868.
[50]   Hunt 1994, pp. 53–54.
[51]   Kelvin, Glasgow, J25, Jenkin to Thomson, 3 January 1860.

Sloan believes that the decision to combine was based on two factors, first 'the potential synergy of combining Thomson's work with the mirror galvanometer and Jenkin's work with the electrical characteristics of long submarine cables', and secondly the possibility of jointly producing an improved sending device.[52] This implies that the central aim of partnership was innovation, and that profits, while eagerly anticipated and catered for in the formal agreement, were not the driving force in Thomson and Jenkin's relationship. The idea that Thomson took on Jenkin because of the latter's

Figure 4.2:   Coiling the cable (W.H. Russell *The Atlantic Telegraph*)

close relationship with members of the Red Sea scheme, is not supported by evidence.[53] In 1868 Varley presented an account of their work to Sir Richard Glass of the Anglo-American Telegraph Company which claimed that far from realising profit, the partners had incurred unremunerated expenses amounting to thousands of pounds. In Varley's view their achievement was an unrivalled technical and scientific contribution to submarine telegraphy:

> I was the first to discover and Thomson the first to work out mathematically the laws of retardation. I was the first to test for the position of faults in submarine cables in 1864 and subsequently Thomson, Jenkin and I have continually led the way in the development of the theory and practice of testing and working cables. The whole question of achieving high speeds has been gradually developed by our three selves alone – *vide*: my paper to the British Association in 1864 and

---

[52]   Sloan 1996, p. 134.
[53]   This was suggested by Smith and Wise: 1989, p. 701.

Thomson's, Jenkin's and mine in the government report 1859 and elsewhere.[54]

While his estimate of the costs of experimenting may have been accurate, Varley's other claims have been disputed. It is suspected that he tended to exaggerate his own contribution to telegraphy, although in this he was not alone, for in every aspect of submarine telegraph innovation different engineers competed for credit. The truth is often elusive, and identifying the origins of methods and systems is often more difficult than tracing the development of an instrument. There are also cases where practitioners working quite independently made simultaneous discoveries, and Varley's

Figure 4.3: Paying out machinery (W.H. Russell *The Atlantic Telegraph*)

contribution to the laws of retardation in cables seems to belong in this category. His brother Samuel recorded in 1859 that he had assisted Varley in experiments on gutta percha covered wires late in 1851, when 'retardation', also referred to as 'induction', was first identified. Samuel Varley noted that Faraday had first discovered the general phenomenon, the capacitance effect, in 1838, and that Werner Siemens was working on similar lines to Varley in Prussia in 1850, although his findings were not known in Britain until 1854.[55]

Varley's assertion that he was the first to test for faults in submarine cables has also been challenged. Jenkin wrote to Sir Charles Bright early in 1865 requesting information about the use of resistance coils for locating faults in cables. Presumably knowing of Varley's tests in 1864, Jenkin sought information in order to establish precedence and determine whether there were patentable methods or instruments. Bright, as one of the pioneers of long

[54] Kelvin, Glasgow, V20, Varley to Thomson, 10 June 1868.

[55] See Samuel Varley, 'On the practical bearing of the theory of electricity in submarine telegraphy, the electrical difficulties in long circuits and the conditions requisite in a cable to insure rapid and certain communication', *Journal of the Society of Arts*, 1 April 1859, especially p. 306. For this reference and other ideas about the Varleys' contribution to this development, our thanks are due to John Varley Jeffery.

submarine cable-laying, could cite the use of techniques similar to those of Varley by several other engineers. Bright himself claimed to have carried out such tests on underground cables in 1851, and hinted at similar work by the Siemens brothers in the 1850s, which had remained unpublished. He added his views on Varley's character:

> So many hallucinations appear to exist in [Varley's] mind on the subject of his inventions that it is only to be [trusted] when he can produce actual printed [memos] of undoubted date. I believe he puts forward a claim at some very early date, that if he had made any experiments at all before a comparatively short period he must either have abandoned them as useless, or have determined to keep his knowledge to himself for both Mr Edwin Clark and Mr Latimer Clark ... were consulting engineers to the Atlantic Company while Mr Varley was employed in a subordinate capacity.[56]

Bright thought that Varley, 'after trying to persuade others ... has succeeded in persuading himself that he thought of the use of resistance coils at some very early date' and urged Jenkin to be cautious about Varley, exhorting him 'to accept as evidence in such matters nothing but positive record'. Yet Varley had already published, in 1859, the method of fault location which continued to be known as the 'Varley Loop Test' into the twentieth century and his claim to originality in this cannot be lightly dismissed.[57]

Varley's engineering achievements were considerable, and when the three formally joined forces in 1865, Thomson, Varley and Jenkin were considered the leading experts in submarine telegraphy.[58] Varley (1828–83), member of a London family of engineers and artists, had begun his career with the Electric Telegraph Company in 1846, rising to become its Chief Engineer in 1861.[59] At the age of eighteen he was placed in charge of an underground cable in London which was plagued by bad joints.[60] This work sparked a longstanding interest in tracing the position of faults in submerged cables, although Varley's many innovations also included improvements to overhead telegraphs. He was the first to measure the speed of current in submarine cables, and produced a machine to magnify electrical charges which formed

---

[56] Institution of Electrical Engineers, Letter Book 1, Sir Charles Bright to Jenkin, 25 February 1865.

[57] 'On some of the Methods adopted for ascertaining the Locality and Nature of Defects in Telegraphic Conductors', *Transactions of the Institution of Civil Engineers* for 1859, pp. 252–55.

[58] Smith and Wise 1989, p. 702.

[59] Jeffery 1997, pp. 263–79; and see Varley's evidence to the Galton committee, especially pp. 148–9. Varley's life is described in an obituary in *The Electrician*, 8 September 1883, pp. 397–8.

[60] Hunt 1994, pp. 50–51.

the basis for Thomson's multiplier.[61] Varley met Thomson in Ireland while working on the 1858 Atlantic cable project, and like Jenkin had been in technical correspondence with him since 1859.[62] During the proceedings of the Galton committee the testimonies of the three combined to undermine the credibility of the Atlantic Telegraph Company's first electrician, the surgeon-turned-engineer Wildman Whitehouse.[63] Varley later said that his objective in joining Thomson and Jenkin was so that the three could act 'in concert rather than in opposition' to secure some reward for their years of work. But they were soon victims of their own technical achievement, for 'when the [1866 Atlantic] line was laid everything went off so successfully that nobody seemed to remember there had been any difficulties to surmount'.[64]

The problems which the partners confronted were above all electrical. Senders, cables and receivers worked as a system, and while increases in speed of signalling were obtained by paying attention to each as components, the importance of mutual interaction had to be recognised. In the 1850s Thomson and Varley had urged careful quality control in the manufacture of conducting core and insulating envelope; the result was enhanced performance of both components, and a much improved cable. Thomson and Varley took out a patent in 1865, for a new kind of sending key which automatically sent reversing currents after each action of the operator in order to sharpen the signal for the receiver, a technique referred to as 'curbing'.[65] For this to be effective a receiver was required which would echo its characteristics and complement it in operation. Throughout the 1860s Thomson sought improvements to his galvanometers; high sensitivity to electric currents, low mechanical inertia and small restoring forces were essential features. He found a solution in the siphon recorder, one of his finest achievements.[66] A coil suspended in the field of a powerful magnet moved under the influence of very small currents.[67] The coil was weakly coupled to a very light, very thin glass siphon, which had one end dipped in an ink reservoir while the other wrote on moving paper tape. In later versions of the recorder a high electric field applied between the siphon and the paper caused ink to be ejected in short spurts, resulting in a trace consisting of a dotted wavy line. When matched to automatic or semi-automatic senders the siphon recorder enabled rapid transmission with the further advantage of a permanent and accurate record on paper tape. (See Figure 4.4) The message could be transcribed by just one

---

[61]   *Electricians' Directory*, 1884, pp. xii–xiii.
[62]   Sloan 1996, p. 131.
[63]   Smith and Wise 1989, pp. 676–77, 702.
[64]   Kelvin, Glasgow, V20, Varley to Sir Richard Glass, 10 June 1868.
[65]   Sloan 1996, p. 217. The patent was number 1865/1784.
[66]   While the instrument seems to have worked well in the laboratory and under close control elsewhere, where conditions were extreme it required modification.
[67]   Both permanent and electromagnets were used.

clerk, saving on the two required by the mirror galvanometer, at up to sixty words a minute.[68]

Jenkin also continued work on mechanical problems associated with deep-sea cables and other applications, developing an interest in control

Figure 4.4: Details from siphon recorder (Prescott, 1877, p. 559)

systems which stemmed from his involvement with cable-laying in the late 1850s. Bright and others on the failed Atlantic expeditions of 1857 and 1858 had realised the importance of improving techniques to regulate the strain on telegraph cables during laying. It was essential to match the speed of the ship with the rate at which the cable was paid out, with tension maintained well below the breaking strain of the cables. If the forward speed of the ship was too high in relation to the speed of the cable then excessive tension could snap the cable.[69] If the speed of the ship was low in relation to the speed of the cable then an excessive length of cable would be used. The size of the ocean swell was a further complicating factor. In 1866 Jenkin took out sole patent rights on a self-adjusting brake strap which controlled the retarding force when laying or retrieving cables. He had also, in 1865, patented improved machinery for manufacturing cables.[70] These mechanical ideas he carried beyond cable

[68]   Headrick 1991, p. 32.
[69]   This actually occurred on the first attempt across the Atlantic in August 1857.
[70]   Sloan 1996, p. 220; obituary of Jenkin in the *Institution of Civil Engineers Proceedings*, pp. 371–2; patents 1865/ 2155, 'Improvements in Machinery to be used in the Manufacture of Telegraph Cables', and 1866/ 1218, 'Improvements in Apparatus for Winding in Telegraphic Cables, applicable also when Winding in Other Ropes or Chains'.

laying: in 1860 he wrote to Thomson on steam engine governors, while apparatus produced in 1863 to determine the ohm included a governor of his design.[71] The same ideas were applied in the 1860s to the control of Jenkin's sender, and to the design and construction of a constant flow valve.[72] In the 1880s Jenkin supplied a wide variety of designs for the control of his telpherage system.[73]

Thomson and Jenkin's joint work during the 1860s demonstrates a continuing cross-fertilisation between electrical theory and telegraphic practice. In 1862 the pair were working separately – though in close communication – to improve various instruments.

> The galvanometers were made in Glasgow under the personal superintendence of Professor Thomson; the resistance coils, switches and keys in London, under the personal superintendence of Mr Fleeming Jenkin, the unit of resistance being the 'B.A.' (British Association). The keys, switches, &c., looked very pretty, and were indeed *multum in parvo*; all were mounted on the bottom of a vulcanite case, with a cup at each corner to hold chloride of calcium; the whole was covered by a glass lid through which keys to manipulate switches, &c., could be inserted. But in this arrangement the connections interfered with each other, so it had to be abandoned. It was while engaged, on March 3rd, 1862, in endeavouring to trace the cause of so much faulty insulation, that Mr Jenkin noticed that Ohm's law does not hold good after you get a certain tension.[74]

Instruments for research, teaching and commercial use were made for Thomson in Glasgow by his long-standing associate James White, optician and philosophical instrument maker, who was able to convert Thomson's ideas into practical shape as well as suggest improvements.[75]

Despite their growing technological and financial successes after 1865, relationships between the three partners were not easy. Thomson tended to arrogance, Varley could be hasty or lapse into illness when placed under stress, and Jenkin, despite his reputation for directness, was cast as peacemaker.[76] An advantage which he brought to this role was that he never took offence; he said

---

[71]   Kelvin, Glasgow, J38, 8 August 1860, and J39, n. d., Jenkin to Thomson. The apparatus is discussed in S. Bennett, *A history of control engineering, 1800–1930* (London, 1986), pp. 64–71.

[72]   See C. Singer et al., *A History of Technology: the late nineteenth century*, V, p. 536.

[73]   See Chapter 6 below.

[74]   Smith 1974 [1891], pp. 99–100.

[75]   Sloan 1996, p. 133; Clarke et al., 1989, ch. 21.

[76]   It has been suggested that Varley was lethargic and willing to allow colleagues to work while claiming credit for himself. These characteristics were largely responsible for a rift with his brother Samuel, who was also a telegraph engineer of some talent and originality. For these details we are grateful to John Varley Jeffery.

once that he could not remember a single malicious act done to himself: 'In fact it is rather awkward when I have to say the Lord's Prayer. I have nobody's trespasses to forgive'.[77] When the first version of the siphon recorder was produced in 1867 and the partners disagreed about how best to market the instrument, Jenkin sided with Thomson against Varley, to whom he wrote: 'You and I are not responsible for the excellence of Thomson's invention'.[78] His deep respect for Thomson's ability made easier the job of restoring harmony, though Jenkin's transparent admiration probably increased Thomson's inclination to condescend to his partners. This tendency shows in Thomson's testimonial to the Royal Society in support of Varley's election to F.R.S. in 1871. While praising Varley's technical achievements:

> I do think that the beautiful and ingenious electric experiments containing as they do some very decided and important discoveries of new and previously unthought of electrolytic properties which he has recently communicated to the Royal Society are [sic] very positive merits not inferior in weight to those generally considered as ample qualification for a Fellow.[79]

Thomson's support could be interpreted as less than whole-hearted, for he also noted:

> I have known him since 1858 ... and I have found him most open and trustworthy in all his doings and sayings. He may often be and indeed too often is, hasty in rushing to conclusions and that I believe his chief faults in scientific and philosophical judgement are due to a certain candour and generosity of disposition by which he is led to be not so sceptical as he ought to be. He certainly has very decided genius and with a better early education would have been a more accomplished Fellow of the Royal Society. All things considered I do not think that his 'mesmerism' or 'spiritualism' ought to be fatal to his being elected.[80]

Thomson produced a similarly equivocal view of Jenkin in his 'Note on the Contributions of Fleeming Jenkin to Electrical and Engineering Science', which had to be censored before it could appear as Appendix I to Jenkin's collected papers.[81] Sloan has noted the 'dry recitation', the impersonal tones of Thomson's words about Jenkin.[82] The obituary of Jenkin in the *Institution of*

---

[77] RLS, p. cxxix.
[78] Kelvin, Glasgow, J71, Jenkin to Thomson, 3 July 1867.
[79] This quotation was supplied by John Varley Jeffery.
[80] Quoted by Jeffery 1997, p. 268.
[81] Yale University Library, Beinecke collection, 4390, Colvin to Stevenson, 13 September 1887.
[82] Norton Q. Sloan, personal communication, 19 April 1997.

Civil Engineers Proceedings claiming 'a close personal friendship' with Thomson suggests a greater affection than actually existed, for although the pair had spent a holiday together on Arran, their relationship was essentially a professional one.[83] Jenkin's comment to his wife in 1869 – 'I *do* like Thomson'[84] – is a strange one about a man he had known well for more than ten years, and who had been his partner for eight, if Thomson were indeed considered a friend.

Jenkin, perhaps with Bright's warning in mind, was never entirely at ease with Varley. He wrote to Thomson that he did not trust 'Varley's discretion'.[85] When Varley went to New York to take out patents on the partners' instruments, the strain of trying to produce clear specifications, which in the American system had to be intelligible to a jury of 'twelve common men', made him unwell:

> The attempt to reduce the 1858 patent to a satisfactory state has made me quite ill. I came here because my brain was overtaxed and overexcited and I hope you will not presume with anything more than is absolutely necessary for I am still very excitable ...[86]

Although Varley did take on business responsibilities within the partnership, much of the commercial work fell to Jenkin. Varley's emotional fragility, Thomson's towering scientific reputation and also his distance from London, meant Jenkin was best placed for this role. But despite his admiration for Thomson, Jenkin was never subservient. Even with Thomson, he could be blunt to the point of abruptness: 'If you are publishing anything about forces while laying cables without reading Brook and Longridge you are *wrong*', he told his partner in 1866.[87] Yet he was quick to apologise, as in 1865 when he begged pardon for having been 'cantankerous',[88] and he acted as mediator, within the partnership and beyond. In August 1868, after a serious misunderstanding with Thomson, Jenkin tried hard to smooth over the difficulties:

> I now think only that you wrote in such haste as not to see how much pain your comments were likely to give. It does not much matter how agreements are construed, nor that we should take different views as to

---

[83]   Obituary in the *Institution of Civil Engineers Proceedings*, p. 366; Kelvin, Glasgow, J65, Jenkin to Thomson, 23 May 1867.

[84]   RLS, p. cx.

[85]   Kelvin, Glasgow, J78, Jenkin to Thomson, 21 August 1867.

[86]   Kelvin, Cambridge, V1, Varley to Thomson, 29 November 1867.

[87]   Kelvin, Glasgow, J56, Jenkin to Thomson, 12 February 1866.

[88]   Kelvin, Glasgow, J53, Jenkin to Thomson, 13 July 1865.

their construction, but it does matter entirely that we should all think well of one another. I shall as you say consider certain things unwritten ...[89]

He emphasised that they must avoid becoming 'rivals instead of partners'. Mutual trust and flexible business arrangements would enable each partner to follow his own innovative instincts:

> An inventor always takes a different view as to the value of his invention from that taken by his neighbours and we should each in turn be annoyed by cold water thrown on our projects. Then if instruments are paid for without previous sanction, who can set a limit to what one of us may choose to spend and if no instrument is to be made without the sanction of the other partners how horribly fettering such an arrangement would be. It was the great if not insuperable difficulty of settling all these points which made us agree to the rough and ready mode of leaving each man to bear the expense of his own speculations ... I doubt whether one or the other mode of settling the expenses will make £20 difference to any of us, and we shall all have enjoyed perfect liberty of action and have also been free from the idea that any one was spending money wantonly or uselessly.[90]

Previous attempts to account closely for spending on experiments and trials of instruments had, Jenkin implied, caused intolerable frustration and complication. Whatever the personal tensions between them, in financial matters the partners inclined towards pragmatism, an approach which proved to be very successful. Each man pursued his own 'speculations' without constant reference to the others. Years later, after Jenkin's death, which followed closely upon that of Varley, this flexibility in partnership affairs caused problems for Jenkin's son in winding up their accounts. Austin Jenkin, a barrister, complained of having 'considerable difficulty' in producing a final settlement as he could find no trace of an account book covering transactions before 1869.[91] Fleeming Jenkin's comment in 1865 to Thomson that 'accounts are a mystery and not the least like mathematics' may not have been entirely in jest.[92]

Yet in their dealings with others, the partners did not lack method. From the time of Varley's joining, much of the correspondence between them related to matters of commerce. According to an obituarist in 1885, 'their patents covered the only methods by which long submarine cables could, or still can,

---

[89] Kelvin, Glasgow, J88, Jenkin to Thomson, 13 August 1868.
[90] Kelvin, Glasgow, J88, Jenkin to Thomson, 13 August 1868.
[91] Kelvin, Cambridge, J21, Austin Jenkin to Thomson, 27 October [1887?]. Sloan discovered that routine administrative matters such as patent renewals were handled for the partners by Jenkin's other associate, H.C. Forde: Sloan 1996, p. 192.
[92] Kelvin, Glasgow, J51, Jenkin to Thomson, 16 February 1865.

be practically worked, their apparatus was used by all the great companies, and the partnership proved extremely profitable'.[93] This attainment was as much a tribute to Jenkin's business enterprise as to Thomson's inventive talent. As commercial manager, Jenkin 'showed himself to be not less a man of business than a scientific theorist and a practical engineer, and his excellent tact and thorough fairness secured for the patents a freedom from litigation remarkable in so valuable a monopoly'.[94] By the time of Jenkin's death, the partnership was regarded as 'synonymous with the progress of submarine telegraphy'.[95] It ended only with the sudden deaths first of Varley in September 1883, and then in 1885 of Jenkin himself.[96]

Jenkin's commercial acumen, so critical to the partners' success, developed from his earliest encounters with Thomson. Recognising the potential financial rewards of their work, he sought Thomson's guidance on patenting. Along with a paper intended for public delivery, he wrote to Thomson: 'You will see that ... I make a practical suggestion for improving the speed of signalling – if this be new, true and patentable perhaps it had better not be read – but I feel the chances are against the three adjectives being applicable.'[97] Though anxious to earn money, Jenkin was uncertain of the commercial value of his discoveries and concerned to act with propriety. 'I know patents are sadly unscientific things but I have hardly any other mode of claiming my own ideas ...', he wrote, on a plan to patent a mechanical sender and 'various other contrivances relative to submarine cables' in 1859.[98] Later he wrote to Thomson that he did not wish 'to be or to seem greedy – it is so difficult to settle in one's own mind what is fair'.[99] This problem of setting and receiving a fair price was to become a recurring one for the partnership.

Jenkin's first patent was taken jointly with Thomson in August 1860. Entitled 'Improvements in the Means of Telegraphic Communication', its main feature was the system of curb-sending, speeding the rate of transmission by sharpening the signal, which was achieved through rapid reversals in polarity.[100] A month later, Jenkin was trying to arrange French patents.[101] He was coming to understand the wider economic implications of their work: in

---

[93] Obituary in *Institution of Civil Engineers Proceedings*, p. 367.

[94] Obituary in *Institution of Civil Engineers Proceedings*, p. 367.

[95] 'Professor Fleeming Jenkin, LL.D., F.R.S.', obituary in *Nature*, XXXII, 18 June 1885.

[96] *Electricians' Directory*, 1884, p. xii. Varley left £46,000, the proceeds of his career in engineering: Jeffery 1997, p. 274.

[97] Kelvin, Glasgow, J17, Jenkin to Thomson, 9 September 1859.

[98] Kelvin, Glasgow, J24, Jenkin to Thomson, 24 September 1859.

[99] Kelvin, Glasgow, J29, Jenkin to Thomson, 13 January 1860.

[100] Sloan 1996, p. 217. The patent was 1860/2047; see obituary in *Institution of Civil Engineers Proceedings*, pp. 366–7.

[101] Kelvin, Glasgow, J41, Jenkin to Thomson, 7 September 1860.

February 1861, while many still doubted even that an Atlantic cable could be laid successfully, Jenkin was planning how it would work, writing to Thomson that 'the use of our patents would diminish the cost of an Atlantic line seventy-five per cent'.[102] He also considered how to make best use of patents, distinguishing between instruments for use by fellow engineers, which however novel were not worth the trouble and expense of patent protection, and signalling devices, which he identified as amply repaying the investment in a patent:

> I am quite of opinion that it is undesirable to patent your improvement on Electrometers. I quite believe that the instrument will be as useful and as renowned as the Marine Galvanometer; but then what has been made by the Marine? You may think this bad management on our part, but I think this is not so. Engineers cannot afford to pay large royalties (hundreds or thousands) and they are ingenious enough to devise some dodge which will do without infringing patents. Messrs Siemens never uses and never will use a Marine Galvanometer and any attempt to coerce say Bright and Clark to pay large royalties on the Marine Galvr. would simply drive them into Siemens' system, though not so good as yours, and so it would be with Electrometers I think. The signalling instruments are a wholly different case – if they really are better than other people's, they will earn for the Atlantic Co. many pounds for every pound they earn for us – but no testing instrument will do this, at least not obviously.[103]

This he wrote to Thomson in 1865, before the successful completion of the Atlantic cable. Jenkin realised that the selling point of their improved instruments was the economic advantages to telegraph companies, which had to be demonstrated in order to convince the companies to pay a premium for them. Companies would gain from increases in signalling speeds, which meant that lines could carry more traffic, and also through improvements in the strength of signals so that fewer relay stations and staff were needed.[104]

Yet Thomson, Varley and Jenkin, though 'leading experts' in submarine telegraphy, did not benefit immediately from the success in 1866 of the fourth attempt to lay an Atlantic cable. It took a bitter and protracted legal battle to deliver the financial rewards to which Jenkin had looked forward for so long. The main reason for this setback was changes in the structure and management of the companies involved in the Atlantic scheme, with a shift in control towards financiers and cable-makers. Cyrus Field, the New Yorker whose vision and energy had kept the scheme alive when all around him despaired,

---

[102] Kelvin, Glasgow, J25, Jenkin to Thomson, 12 February 1861.
[103] Kelvin, Glasgow, J54, Jenkin to Thomson, 17 November 1865.
[104] See Kelvin, Glasgow, J31 and J34, Jenkin to Thomson, 20 January and 28 February 1860.

was still closely involved, but power had passed to men of less vision whose driving motivation was financial. Varley's observation that the 1866 cable was so successful that straightaway 'nobody seemed to remember there had been any difficulties to surmount' is pertinent. Before 1866, the Atlantic companies could not pay; afterwards, they would not. After a decade of frustration, the big investors looked forward to collecting dividends, and contemplated the launch of further profitable long-distance undersea cable companies. At last there was genuine and well-founded confidence in the technology, but the engineers responsible for developing it, those who had believed all along that the problems were soluble, began to be seen as expendable.

As far as Thomson, Varley and Jenkin were concerned, at the centre of their immediate difficulty was the restructuring of the Atlantic Telegraph Company. They had signed an agreement in July 1865 which gave the company exclusive rights on any Atlantic route to patents taken out by Varley in 1856, 1860 and 1862 for improvements in testing and signalling; to Thomson's patent of 1858 which contained twenty-two parts, including the mirror galvanometer; and to the Jenkin-Thomson patent of 1860.[105] The contract also included a curb key by Thomson and Varley in process of patenting,[106] along with any future patents which the partners might obtain. Payment, crucially, was tied to the net earnings of the company, of which Thomson, Varley and Jenkin would receive three per cent.[107]

After the failure of the attempt to lay a cable between Ireland and Newfoundland in August 1865 the Atlantic Telegraph Company underwent a crisis. There was a certain irony in this, as for the first time there was real confidence that the project would soon succeed. Despite the unprecedented belief in the technology, there had to be further delay while another cable was made. It was planned to lay it during the short summer laying season of 1866. The directors of the Atlantic Telegraph Company set about raising £600,000 through the issue of twelve per cent preference shares, and were assured that the cable manufacturers, the Telegraph Construction and Maintenance Company Ltd (Telcon), would take on the risk themselves if necessary, accepting payment in company stock if public subscriptions did not meet the amount required.[108] Telcon started to make the cable, and there seemed little doubt that the project would shortly succeed.

---

[105] PRO, C16/533/T29 36834; Sloan 1996, p. 216.
[106] Patent 1865/1784: Sloan 1996, p. 217.
[107] PRO, C16/533/T29 36834.
[108] Field 1893, pp. 296–98. The Telegraph Construction and Maintenance Company Ltd. had been formed in 1864 by a merger of Glass and Elliot and the Gutta Percha Company.

At Christmas 1865, Cyrus Field was shocked to discover that the new share issue was illegal. The Atlantic Company had been incorporated by private Act of Parliament without provision to increase its capital. The parliamentary timetable could not accommodate an amendment to the Act before the summer recess. After rapid consultations with business and legal advisers, a way out of the dilemma was found. A new company, the Anglo-American Telegraph Company Ltd., was set up to raise the required money and act as agent in constructing and working the line. Because of the urgency, Field brought in new large investors and advisers, notably Daniel Gooch M.P., chairman of the Great Western Railway who was also owner of the *Great Eastern*, which had laid the 1865 cable and was to lay that of 1866.[109] There was a central role in the new company for representatives of the Telegraph Construction and Maintenance Company, including Sir John Pender and Sir Richard Atwood Glass, so that the cable-makers enjoyed a closer association with the Anglo-Americanthan they had ever had with the Atlantic Telegraph Company.

In summer 1866 the Atlantic scheme was successful at last, with a new cable laid, and the recovery and completion of the abandoned 1865 cable. Almost immediately the boards of the Atlantic company and the Anglo-American fell into dispute about operating issues such as rates for telegrams, and also over the division of profits. Atlantic investors who had stayed with the company during many years of failure felt that they were losing out to Anglo members whose involvement had started only when the Atlantic telegraph was a near-certainty. The Atlantic directors tried to exercise their option of taking over the Anglo-American company, but were defeated and forced to resign. The formal conclusion to this struggle came in 1870, when the Atlantic company was absorbed by its rival.[110]

The Thomson, Varley and Jenkin partnership was a casualty of these events. Their agreement with the Atlantic Telegraph Company became clouded in ambiguity, as the cable lines were run by the Anglo-American and financial arrangements between the companies remained unclear. Additionally the partners had lost personal influence with the directors, as many original board members had dropped out after suffering large losses, or had given way to the new financiers. Thomson himself had served as a director of the Atlantic company from 1856, but resigned in 1859 when the scheme was in abeyance and business activities were diverting him from scientific work.[111] Jenkin could

[109] For brief details of Gooch's career, see Geoffrey Channon, 'Sir Daniel Gooch (1816–89), railway engineer and executive', in Jeremy (ed.) 1984, II, pp. 597–603. The *Great Eastern*, the only ship in the world large enough to carry the entire cable, had been refitted as a cable ship after a brief unsuccessful career as a passenger vessel.
[110] New York Public Library, Wheeler 4719 and 4616.
[111] Sloan 1996, pp. 146, 190.

see that the partners were losing influence with the company: 'I am becoming more and more convinced that unless one has an advocate at the Board one's interests are never attended to'.[112] Varley and Thomson, a long-standing advisor whose work for the company had been unpaid before 1866, were on good terms with Sir Richard Glass, first chairman of the Anglo and joint managing director of Telcon, but Glass was forced by ill-health to retire in March 1867.[113] His replacement at Telcon, Captain Sherard Osborn, apparently less sympathetic to the engineers, immediately wrote a chilly letter to Thomson. The board had 'decided that the staff of consulting electricians whose services had been engaged in connection with the Atlantic Telegraph Cable contract should be reduced', Thomson's annual salary of £500 would cease, and in its place a retainer of £100 as Electrical Counsel was offered.[114] In equally icy tones, Thomson replied that he could not undertake to supervise 'such varied and progressive scientific operations ... on such terms'. For Glass, he had been prepared to do this work and decline all other offers, but felt that the rest of the board did not appreciate the understanding between himself and their former managing director.[115] This exchange marks a low point in relations between the partners and the Atlantic telegraph and cable companies associated with Pender and Gooch. Efforts to contact the Anglo directors by letter or in person were rebuffed.[116]

It was imperative that the financial wrangle be resolved. The partners were not receiving payment, and the Atlantic Company, bound by contract to pay the Anglo-American an annual fee of £125,000 after working expenses were settled, was not expected to show a profit or pay a dividend in 1867.[117] Moreover, the 1865 agreement prevented the partners licensing their instruments to any other transatlantic telegraph. In March 1868, Thomson, Varley and Jenkin filed a legal action against the Atlantic Telegraph Company, in which they claimed that in addition to £1054 which the company had offered, they should also be paid three per cent of the £125,000. The Atlantic company's secretary, George Saward, formally responded in May 1868 by renewing an offer of arbitration. Sir Richard Glass, in February 1868 restored to health and back in business as chairman of the Anglo-American company, was involved in the efforts to settle.[118] Varley, assisted by Jenkin and their solicitor, Pritchard, drafted a long letter to Glass in June 1868 which detailed

---

[112] Kelvin, Glasgow, J55, Jenkin to Thomson, 10 January 1866.
[113] Lawford and Nicholson (eds) 1950, p. 65.
[114] Kelvin, Glasgow, O3, Osborn to Thomson, 23 March 1867. The paid consultancy had been recently arranged.
[115] Kelvin, Glasgow, O4, draft of Thomson to Osborn, 1 April 1867.
[116] Kelvin, Glasgow, J80, Jenkin to Thomson, 5 September 1867.
[117] Smith and Wise 1989, p. 703.
[118] New York Public Library, Wheeler 4598.

the background to the case and defined the ways in which telegraph companies had benefited from the partners' work.[119]

Varley was eager to stress that he sought a 'friendly settlement' and that the legal action had been taken only when all other means had been exhausted. Thomson, when a director, had experimented at his own expense, while the original electrician (Wildman Whitehouse) and engineer (Sir Charles Bright) of the company, whose discoveries had been of little practical use according to Varley, received thousands of pounds of the company's money for experimentation and were paid a quarter share each of £75,000 in company stock. Thomson was called at the last minute to stand in – unpaid – on the 1858 expedition for Whitehouse, who was ill. Varley was later asked to investigate the reasons for that cable's failure, and after working sixteen hours every day for two weeks had been paid fifty guineas, which included his expenses. He then served as electrician to the Atlantic company 'gratuitously' from 1859 until 1864, 'the company being in extremis'. He paid even his own railway, hotel and cab fares during this time, and in 1866 covered the costs of training staff to work the new line. Varley estimated that the partners had spent £3000 of their own money on 'bringing our inventions to perfection', yet had been willing to take a risk on the success of the cable by accepting payment related to the Atlantic Telegraph Company's net receipts. The company had estimated in 1865 that Thomson, Varley and Jenkin would receive around £15,000 in the first year. The agreement prevented the partners from contracting with other schemes 'which with our support might have become dangerous rivals'. Their instruments enabled the Atlantic cables to carry six times the traffic they might otherwise have done.

> The possession of [the 1865] agreement allowed the Company to promise confidently to the public a rate of transmission, which had been persistently deemed impossible by a large section of the scientific world ... It is not too much to say that but for our promises as to the transmitting power of the proposed cable, the capital of the undertaking could hardly have been found.

Varley also suggested that he and his partners could offer still greater speed and efficiency to the working of the Atlantic telegraph:

> If the Company had desired it – and if the Company still choose to employ other parts of our inventions, messages can be sent with greater rapidity and with greater accuracy and by an invention patented last year they can be recorded.

---

[119] Kelvin, Glasgow, V20, Varley to Glass, 10 June 1868. For a transcription of this letter and ideas about its significance, we are grateful to Norton Sloan. Jenkin's role in helping draft the letter is mentioned in, Glasgow, J85, Jenkin to Thomson, 27 July 1868.

In round terms, the company was offering £3000 to settle the claim up to May 1868, while the partners held out for £10,000. Varley, who believed that they had 'the strongest moral claim' to £10,000, suggested to Glass that in view of the uncertainties of the legal process they would accept £7000, but were prepared to go to court otherwise. As the company did not insist upon exclusive rights to the instruments, and did not wish to use Thomson's newly patented siphon recorder, the partners would accept an annual payment of £3000 for the following ten years.

Although these proposals were accepted by the company in July 1868, arguments and backtracking continued. The issue was complicated by the Anglo-American's take-over of the defeated Atlantic Telegraph Company, and by Glass's recurring illness which caused him to stand down as chairman of the Anglo.[120] The company had outraged Thomson, Varley and Jenkin by attempting to change the agreement, reducing its period from ten years to eight and a half, the length of time for which the patent on Varley's condenser still had to run. In July 1869 the agreed £7000 was finally paid, and annual payments for use of the partners' instruments became routine.[121] Two clauses in the final contract were crucial in persuading the partners to accept a lower figure than they had first claimed. In place of granting exclusive Atlantic rights on their instruments as agreed in 1865, the partners now undertook merely that they would not give any rival Atlantic telegraph companies preferential terms. Thomson, Varley and Jenkin were also given a privileged relationship with the Atlantic and Anglo companies: they were to be allowed time to match any better signalling devices submitted to the telegraph companies by rival electricians, effectively giving their instruments a ten-year monopoly on the lines.[122]

The settlement enabled open negotiation with other companies planning transatlantic telegraphs. In July 1867 Jenkin had reported to Thomson an approach by promoters of another Atlantic cable.

> I answered that I could give every assistance except that of expediting the transmission of messages. I presume I may report on other systems as I have also been asked to do but that Atlantic agreement places one in a most awkward position and may be the means of not only fettering one directly but of preventing the obtaining of an appointment as Engineer to an independent company.[123]

---

120 New York Public Library, Wheeler 4599.
121 Sloan 1996, pp. 155–59. The draft agreement is at Kelvin, Glasgow, J92.
122 Smith and Wise 1989, p. 704.
123 Kelvin, Glasgow, J71, Jenkin to George Saward, 3 July 1867.

A year later Jenkin was approached by Sir James Anderson, master of the *Great Eastern* on her cable-laying voyages of 1865 and 1866, who was promoting two new telegraph companies, an Anglo-Indian line and the French Atlantic cable. Anderson wanted to secure the partners' services and patents, and offered £1500 a year from the Indian company, and £3000 for the Atlantic project.[124] This last offer explains why Thomson, Varley and Jenkin were so eager for a rapid conclusion to their dispute with the Atlantic Telegraph Company, even if it meant conceding part of their original financial claim. The high payment offered by Anderson for their services on an Atlantic line reflects the value of their French and American patents. In the event, the Indian line came to nothing, but Jenkin rapidly reached agreement with the main promoters of the French Atlantic cable, Julius Reuter of the eponymous news agency, and Baron Emile d'Erlanger, a Parisian banker. Experience had taught him to be 'strongly in favour of agreeing at once before the Directors have become numerous and have time to raise difficulties, look into schedules of patents etc. etc.'[125] He subsequently wrote to Thomson:

> I am happy to tell you that Erlanger and Reuter today signed the agreement for the use of our instruments £3000 per annum for ten years subject to same condition as the inchoate Anglo American agreement. Also £3000 for you and Varley as Consulting Engineers and Electricians. I expect you will pat me on the back for this.[126]

Jenkin was doubtful that the Indian scheme was proceeding, but had more faith in Reuter and Erlanger: 'They are sure to carry it out. Clark and we are Engineers, you and Varley Consulting Electricians and Engineers ... I am confident the capital is secured, Reuter and Erlanger being both men of very high financial position ...'[127] He expressed hopes that the case against the Atlantic board would be settled before the Reuter agreement became public. In the event Jenkin's confidence in Reuter proved well founded and the Brest-St Pierre cable was actually laid before the Atlantic case finally concluded.

In a very few years, transatlantic telegraphy had turned from a high risk venture into a routine procedure – at least in the minds of the public, but increasingly also in the view of large investors. The French cable was to be the third line across the Atlantic, and although less risky than its predecessors it still presented challenges. It was not, strictly, a French cable, for the company promoting it, the Société du Cable Transatlantique Francais, was a limited liability company formed in August 1868 in London with a capital of £1.2

---

[124] Kelvin, Glasgow, J83, Jenkin to Thomson, 14 July 1868.
[125] Kelvin, Glasgow, J83, Jenkin to Thomson, 14 July 1868.
[126] Kelvin, Glasgow, J84, Jenkin to Thomson, 21 July 1868.
[127] Kelvin, Glasgow, J85, Jenkin to Thomson, 27 July 1868.

million, and commonly known as the French Atlantic Company. The length of
the line, from Brest to St Pierre, a French island south of Newfoundland, and
thence to Duxbury, Massachusetts, was over 3000 nautical miles, almost
double the length of the 1865 and 1866 cables between Valentia in the west of
Ireland and Heart's Content, Newfoundland.[128] In 1868 Erlanger and Reuter
had negotiated with the French emperor an exclusive twenty-year concession
to lay and work a cable between France and the United States, for which they
lodged 'caution money' of £20,000. They had also bought for £10,000 a pre-
existing similar concession in the State of New York.[129] These arrangements
were conditional on the line being in operation by 1 September 1869.

The French cable project was unprecedented in scale and cost, and also
in its detailed organisation. Contracts were drawn up in July 1868 with Clark,
Forde and Jenkin, to act as consulting engineers and electricians for a fee of
£10,000; with Thomson and Varley, consulting electricians, who would
receive £3000; with Thomson, Jenkin and Varley, to be paid £3000 *per annum*
for ten years for use of their patents; with James Anderson, who would be
general superintendent for a fee of £4000 plus expenses; and with Telcon, who
were to receive £920,000 for making the main section of cable and laying the
whole line.[130] A programme of proceedings, published in London on 1 June
1869, was necessary to co-ordinate the work of the four ships involved. The
Great Eastern carried 2752 miles of cable, weighing 5480 tons; three other
steam ships carried mainly shore end cable, a further 800 miles or 3000 tons.[131]

The French shore end was laid at Fort Minou, ten miles from Brest, on
19 June 1869, and the *Great Eastern* sailed for America two days later.
Thomson and Forde stayed at the French end during the laying, and for a
month's testing afterwards. Sir Samuel Canning and Henry Clifford worked
for Telcon in charge of cable-laying, and Willoughby Smith was Telcon's
electrician on board ship. Varley, Clark and Jenkin were all on board,
representing the telegraph company, along with various assistants. Jenkin had
taken Leslie C. Hill, one of his students, 'my prizeman at University
College'.[132] The ship was under the command of Captain Robert Halpin, and
there were also present his predecessor Sir James Anderson, referred to as
director general of the French company, and Sir Daniel Gooch, chairman of
both the Great Eastern Steamship Co., which owned the ship, and the
Telegraph Construction and Maintenance Co.[133] Gooch was also still a director

---

[128] Bright, 1974 [1898], p. 107; National Maritime Museum, ms 88/078.
[129] PRO, BT31/1418/4105. See also New York Public Library, Wheeler 4605.
[130] New York Public Library, Wheeler 4605; Bright, 1974 [1898], p. 107.
[131] National Maritime Museum, TCM/6/5.
[132] *Illustrated London News*, 10 July 1869; RLS, p. cix; National Maritime Museum, ms 88/078.
[133] See Channon on Sir Daniel Gooch in Jeremy (ed.) 1984, II, p. 600; Wilson (ed.) 1972, p. 111. For Halpin, see Rees 1992.

Figure 4.5: Arrival of the *Great Eastern* off Brest, June 1869
(*Illustrated London News*, 10 July 1869)

of the Anglo-American, supposedly a competitor of the Société du Cable, and
although not a director of the French company, he had a direct interest in their
success as the *Great Eastern's*, Figure 4.5, charter price was £1400 plus
£20,000 in telegraph company shares.[134]

The expedition had been launched with a civic reception in Weymouth
and parties on board, in England and France.[135] Despite the scale of the project,
there was a general air of optimism and the large contingent of experienced
and reputable engineers and electricians added to the expectation of success –
there were twenty-one electricians among the 443 people on board the *Great
Eastern*.[136] Varley described the ship's admirable state of readiness to
Thomson:

> Unlike the 1865 and 1866 expeditions, all the work about the machinery
> is ready and no work going on about the ship of any consequence.
> Everything is before hand and bears unmistakable evidence of Capt.
> Osborne's master mind, Sir S. Canning's ability and Capt. Halpin's
> control of everything ... everything looks like success.[137]

---

[134] Rees 1992, p. 80.
[135] Wilson (ed.) 1972, pp. 153–57.
[136] Wilson (ed.) 1972, p. 168.
[137] Kelvin, Cambridge, V6, Varley to Thomson, 21 June 1869.

Jenkin commented upon the sense of order and confidence in a letter to Anne: 'The look of the thing was that the ship had been spoken to civilly, and had kindly undertaken to do everything that was necessary without any further interference'.[138] He was one engineer among many, and this was less of an adventure than his previous trips had been.

Yet in spite of the careful preparations, the planning, the clear specifications for the cable, the apparent harmony and optimism, all did not go smoothly. The mix of business leaders and various factions of technicians was never likely to be a happy one, and although the cable contract specified rules for laying and testing there remained wide scope for disagreement between the representatives of the telegraph company and those of the contractors.[139] Anderson, chief representative of the company and nominally leader of the expedition, shared a mutual antipathy with Canning, engineer-in-chief for Telcon. Willoughby Smith,, Telcon's chief electrician, disliked and mistrusted Jenkin, and to a lesser extent Varley. While compelled to admit respect for Thomson and his mirror galvanometer, Smith was scathing about his partners:

> [Varley] certainly did not invent [the condenser], and was not the first to apply it in the working of submarine cables. As regards Mr Jenkin, I could not at all see where he came in, unless under the assumption that union is strength ... it seemed to me ... the contractors were entirely in the hands of Messrs. Thomson, Varley, and Jenkin, and that to them they gave great power without the least responsibility.[140]

Access to the testing facilities was the focus of much disagreement. The contract stipulated that testing should be in the hands of the contractors 'but shall be open to the continual inspection of the engineers or their assistants on board ship and on shore'.[141] Clark, Forde and Jenkin had been instructed 'to inspect the electrical tests carried on by the contractors' but also to 'carefully abstain from interfering with the contractors ... about their mode of making such tests' and to 'abstain from making any remarks or expressing any opinions ... as to the propriety or otherwise of their proceedings in regard to the electrical testing, or to the paying out of the cable'.[142] Thomson was to be arbiter in cases of differing interpretation. Thomson, though, was in Brest, and even in the testing room there Forde later complained that he and his assistant had been excluded 'for hours at a time'.[143] On board the *Great Eastern*, Varley

---

[138] RLS, p. cix.
[139] Willoughby Smith's account includes a full 14-page transcript of the cable contract: Smith 1974 [1891], pp. 212–26.
[140] Smith 1974 [1891], p. 230.
[141] Smith 1974 [1891], p. 230.
[142] New York Public Library, Wheeler 4446.
[143] New York Public Library, Wheeler 4608.

was at one point ordered from the testing room, and the engineers were denied access at other times by Willoughby Smith . Smith, protested that 'nothing could be said or done in the test room without one of the Company's staff, notebook in hand, demanding to know what was going on ... frequent complaints were made by me, both to Mr Jenkin and Mr Varley, of the way in which they interfered in the test room during my absence'.[144] He objected at the time taken to carry out official tests when laying was complete, and was affronted that Jenkin and Varley chose to make their own tracings of connections rather than use diagrams provided by Smith's staff. 'I regretted that Mr Latimer Clark was not present, as he was familiar with the behaviour of long cables.'[145] By implication, Varley and Jenkin, and others present, were not.

Gooch had also taken a particular dislike to Jenkin. His aversion probably stemmed from the legal action in which Jenkin was a party against several companies in which Gooch had substantial interests. The especial loathing for Jenkin is explained by differences in background and personality. Jenkin did not give Gooch what he wanted, those things which Smith was willing to provide – deference and simple answers, and where possible good news. Gooch, with a great deal to lose if the project failed, was in a state of high impatience and anxiety. In his telegraph interests, as with his railway companies previously, Gooch had followed a policy of vertical integration. Cable core manufacturing had merged with cable armouring and laying to form the Telcon company in 1864; Gooch also owned the ship and a stake in the telegraph company. When difficulties arose in the first week of the voyage, his immediate response was that 'it cannot be any fault in the cable itself as that is perfect'.[146] His journal during the expedition records the kind of explanations which he preferred – a broken wire, sabotage, or the incompetence of a telegraph clerk.[147] Smith, sensing what Gooch needed, was not only full of respect, but was always first to appear at his cabin door with an uncomplicated answer to any problem.

Jenkin, of course, had no time for sycophancy. The instructions from the Société also bound him to a degree of circumspection. There was a nagging problem with the French cable which showed itself early in the voyage, and which was never fully resolved. Jenkin's considered view was likely to produce a complicated answer. Gooch did not care for thinkers; he had remarked of Canning on an earlier Atlantic telegraph expedition that he 'lost many valuable hours in thinking instead of acting, as was too often the case with him'.[148] On the French trip he complained of 'the persuasions and

---

[144]   Smith 1974 [1891], p. 233.
[145]   Smith 1974 [1891], p. 235.
[146]   Wilson (ed.) 1972, p. 158.
[147]   Wilson (ed.) 1972, pp. 165–66, 169.
[148]   Wilson (ed.) 1972, p. 120.

difficulties made by Canning and Halpin' which had threatened to delay matters and which Gooch had over-ruled.[149] With Thomson, Varley and Jenkin, whom he called 'the French engineers', Gooch could try to bully and hector, but he could not over-rule, for on their approval rested the payment from the Société du Cable to his own company.

When the first main fault showed up, five days into the voyage, Gooch summoned a conference of the Société's engineers at two in the morning.

> I asked the French engineers to give me their opinion frankly, as my desire was to do that which would be best for the success of the enterprise, rather than to consider the interest of the Telegraph Co alone. I regretted I did not get from them any decided opinion as to what was best to be done, but they rather pointed out the consequences to the shares of the French Co if the cable was completed with the slightest fault in it. None of them could give me the slightest idea of where the fault might be; indeed, Mr Jenkins [sic] said it might be a ship had injured the cable in Brest.[150]

After the meeting, Gooch vented his frustration in a further note to himself:

> So little reason do I feel to be satisfied with the French engineers in this consultation that I am determined not again to trouble them. Their whole object, to my mind, was to protect themselves from having offered an opinion rather than to protect and secure the interests of their employers ... would-be wise men, but are fools. One might suppose their selfish object was to procure further employment out of litigation. I will accept [sic] Clarke [sic] from the above opinion, as he has in all cases acted with much more openness, and I think honesty, than the others.[151]

Latimer Clark had managed to stay on better terms by taking care to give information which was simpler and more encouraging, presumably not taking at face value Gooch's request for frankness. Gooch, so anxious about faults in the cable that he could not sleep, remained worried even when the electrical condition of the cable improved, and he maintained his feud with Jenkin, altering the wording of an apparently innocuous telegram from Jenkin to Thomson[152] While still in mid-ocean, Clark tried to reassure Gooch:

> Clark told me that the faults we have cut out were a *million* times larger than the one they believe to be in the cable, and that he thinks for all

---

[149] Wilson (ed.) 1972, p. 156.
[150] Wilson (ed.) 1972, p. 162.
[151] Wilson (ed.) 1972, p. 163.
[152] Wilson (ed.) 1972, p. 164.

practicable purposes, even if the fault exists, the cable will work as well and last as long as a perfect cable.[153]

Clark was also in close contact with Reuter by cable, having been asked to send a report of progress every evening in time for the following morning's newspapers.[154] When the expedition reached shallow water off Newfoundland in the second week of July, it was clear that success had been achieved. Yet this gave Gooch further grounds to criticise Jenkin: 'Jenkins telegraphed today that the cable may now be considered successfully laid. What about his fault?'[155]

On July 10, Anderson had cabled the French directors to say that the line was 'commercially perfect'.[156] It was not until six days later that Clark, Jenkin and Varley confirmed to Thomson that the line was 'in perfect working order'

THE LANDING

Figure 4.6: Landing the cable on the North American mainland, July
1869 (Anon., *The Landing of the French Atlantic Cable at Duxbury,
Mass.*: Boston, Alfred Mudge, 1869)

for all commercial purposes. (See Figure 4.6 in which the landing of the cable is depicted). The engineers considered that the small fault, estimated to be within 400 miles of Brest, was unlikely to worsen with careful usage. Relieved by the long-awaited success, they apparently overlooked the unpleasantness with Smith's staff in the testing room and reported to Thomson that 'ample facilities' had been afforded them to carry out their work.[157] Gooch was less

[153] Wilson (ed.) 1972, p. 171.
[154] See Kelvin, Cambridge, S264 to S276, correspondence between the partners and the Société du Cable Transatlantique Francais Ltd., June and July 1869.
[155] Wilson (ed.) 1972, p. 175.
[156] Wilson (ed.) 1972, p. 176.
[157] Kelvin, Cambridge, S276, Clark and Jenkin to Thomson,16 July 1869.

willing to forgive and forget. Like Willoughby Smith, he felt that Jenkin and Varley's official tests on shore had been overly protracted. Gooch recorded his final thoughts about the electricians:

> We have had most unpractical and unreasonable people to deal with in Jenkins and Varley. Clark is not so bad and without the others I think he would do very well, but I trust my company will never again enter into any contract where Jenkins or Varley are to accompany the expedition.[158]

Clark, Forde and Jenkin later produced for the telegraph company their own report, which detailed some of their criticisms of the contractors. Clark complained that Canning had ignored his request to stop the ship as soon as the first fault was identified; as a result a further thirty-five or forty miles of cable had been paid out before the engineers could properly consider what to do. Jenkin's advice to go back, picking up until the fault was discovered, had been ignored. The contractors wanted assurances that there were no further faults in the unlaid cable, which the telegraph company's electricians could not give. Gooch had been indecisive, knowing that he needed their certificate if Telcon was to be paid. Eventually Varley had assented to the contractors' wish to continue. But it seemed ultimately that the company's engineers had been over-cautious, for Clark, Forde and Jenkin's final conclusion after a month's testing of the cable was that the fault was a tiny one in the gutta percha: 'It is difficult to convey a clear conception of the minuteness of the injury which has occurred'. Even if a thousand times greater, telegraph traffic would not have been interrupted.[159]

The Atlantic crossings of 1866 and 1869 marked a new era in submarine telegraphy, and heralded unprecedented confidence in deep sea cables by businesses and public alike, (see Figure 4.7 for the extent of the network in 1871). Yet the lessons which Thomson, Varley and Jenkin learned from their experiences were not all technical ones. Working arrangements between engineers and contractors, the problem of protocol which had so frustrated Jenkin when he was with Newall, still caused strife unless precisely settled in advance. Despite the new confidence in submarine telegraph technology, there was no reduction in the demands upon electricians, for enormous amounts of money were at stake on the Atlantic expeditions. Above all, Thomson, Varley and Jenkin had learned that it was essential to protect their own position, as it had become apparent that the telegraph companies would not readily compensate them for a decade of unpaid work unless contractually obliged to do so. By carefully managing their British and foreign patents, and because of

---

[158] Wilson (ed.) 1972, p. 182.
[159] New York Public Library, Wheeler 4608.

their persistence in the long struggle against the Atlantic Telegraph Company, in 1869 the partners at last began to see substantial profits.

Besides the annual payment to the partners of £3000 by the Atlantic Telegraph Company, there was a further £3000 from the Société du Cable. Varley was also retained as electrician by the French company with a salary of £400 *per annum*.[160] After the completion of the French cable and the settlement of the Atlantic claim, there were approaches to other telegraph companies, not all of which were productive. Varley had been quick to suggest an agreement

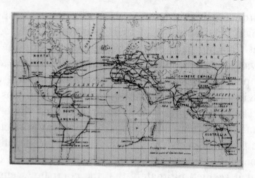

Figure 4.7: The international telegraph network in 1871 (J. Wagstaff Blundell, *The Manual of Submarine Telegraph Companies*: London, Rixon and Arnold, 1871)

with the new Ceylon to Penang company, to act as engineers and electricians for two or three per cent of the gross receipts: 'This will avoid Patents and the Law and yet give us an interest to work the line in the best manner'.[161] This scheme, re-routed to form a line from Madras to Singapore via Penang, went ahead in 1870 under the auspices of the British Indian Extension Telegraph Company Ltd. with Clark and Forde as engineers. Thomson, Varley and Jenkin were not involved in the laying, for although Gooch was not a director, the company was floated by Pender and closely associated with Telcon.[162]

Jenkin's seven-year partnership with Forde had given him valuable experience and contacts. Forde's staff had also eased his burdens by dealing with the routine administration of Thomson, Varley and Jenkin's patents, in liaison with a patent agent, Carpmael.[163] In itself, however, the consultancy

[160] Kelvin, Cambridge, V7, Varley to Thomson, 20 August 1869. The association with the Société lasted into the late 1880s, the partnership receiving several hundred pounds in the year after Jenkin's death: Kelvin, Cambridge, J19, Austin Jenkin to Thomson, 12 March [1887?].

[161] Kelvin, Cambridge, V8, Varley to Thomson, 23 August 1869.

[162] Barty-King 1979, p. 38; Wagstaff Blundell 1871, pp. 25, 52.

[163] Sloan 1996, p. 192.

partnership had not proved lucrative.[164] Consultancy work in submarine telegraphy could be well-paid, but fees varied widely.[165] As well as being seasonal and uncertain, commissions were linked closely to the fortunes of Telcon, the only cable manufacturer with capacity to tender for the biggest contracts. Telcon's monopoly was consolidated by a working agreement with W.T. Henley, lasting from 1864 until 1876, which eliminated Henley as a competitor.[166] Newall continued to make cables but, frozen out of the mainstream and recognising that the battle with Telcon was lost, increasingly diversified into other industries.[167] Yet this commanding position in long cable-making made Telcon reliant upon a few very large contracts. There was no such work available in the immediate aftermath of the 1866 Atlantic triumph and the company experienced two years of 'gloom and anxiety', affecting the whole industry.[168]

Jenkin's family was growing, and these financial fluctuations were a constant worry. In the early years of their marriage, the Jenkins moved several times. At first they had lodgings in Birkenhead. When in the south of England, Jenkin sometimes stayed at Stowting;[169] while away on telegraph voyages, he would leave Anne with her parents in Isleworth. Later Jenkin lodged in London. In 1863 the Jenkins took a cottage outside Claygate, near Esher, surrounded by trees and with a pretty garden, part of which they made into a croquet green. Three sons were born: Austin Fleeming Jenkin, born in London in October 1861; Charles Frewen Jenkin, born in September 1865 at Claygate; and Bernard Maxwell Jenkin, also born at Claygate in 1867. The second son was always called Frewen, and Stevenson later referred to the three as Bruin, Frewen and Stewen: 'I think you will find it has the haunting quality of popular rhymes'.[170] With their new interests in gardening and croquet, frequent parties of friends and neighbours, and their growing family, this rustic life seemed idyllic. Yet there were two clouds hanging over them. After the birth of Frewen, Anne Jenkin was taken ill, and her husband ran miles for the doctor, was soaked by rain and made ill himself. While she recovered, he suffered from 'flying rheumatism varied with settled sciatica', and although generally borne with stoicism, the chronic ailment troubled him in varying degrees for the rest of his life.[171]

[164] See Jenkin's entry in the *Dictionary of National Biography*, XXIX (1892), p. 296.
[165] Kelvin, Glasgow, J58, Jenkin to Thomson, 15 January 1867.
[166] Science Museum Library, HEN 29, 'The early life of W.T. Henley, by himself'.
[167] Russell on Newall, in Jeremy (ed.) 1985, IV, p. 434.
[168] Smith 1974 [1891], p. 209.
[169] See for example his letters in September and December 1860 addressed from Stowting: Kelvin, Glasgow, J42 and J43.
[170] Booth and Mehew (eds) 1995, VIII, p. 117, Stevenson to Anne Jenkin, 18 June 1893.
[171] RLS, p. lxxii.

The other pressing problem was money. Jenkin was ever optimistic, believing that rewards would come for his years of toil. 'I do not feel weak, I do not feel that I shall fail. In many things I have succeeded, and I will in this.'[172] In 1866 he had taken the chair of civil engineering at University College London, which brought in some extra income. But the chair was not endowed, meaning that the professor's salary derived entirely from fees paid by students. This system obliged Jenkin to accept any student to the course, however ill-equipped by education or intellect, and caused him some frustration.[173]

Other aspects of life in London were not to his liking. Jenkin looked back on his work in Duke Street as 'the dusty jolting railway' leading to 'the horrid fusty office with its endless disappointments ..'. While appreciating the idyllic aspects of life at Claygate, he later wrote: 'Now that it is over I am conscious of the weight of anxiety as to money which I bore all the time'.[174] In 1868, he left all this worry behind: his work with Forde began at last to pay; a settlement of the dispute with the Atlantic company was in prospect; and he accepted a salaried chair in engineering, at the University of Edinburgh, where he was to spend the rest of his life.

[172] RLS, p. lxviii.
[173] Samuelson committee, p. 134.
[174] RLS, p. lxxiii.

Chapter 5

A Philosophic Electrician

There were obvious and mundane reasons for Jenkin to be attracted to the Edinburgh chair of engineering. He was relieved to quit a dingy London office associated with so many disappointments and frustrations, in favour of the Scottish capital where he had spent happy childhood years. There was also the security of a regular if modest salary for the new professor of engineering. Jenkin's appointment to the chair in Scotland had been settled in February 1868,[1] during the two year hiatus in long-distance cable enterprises. At that point his business prospects were uncertain, for the French Atlantic cable was not then contemplated, no other challenging projects were in view, and the dispute with the Atlantic company remained unresolved.

Yet the new appointment offered more than security and pleasant surroundings. That Jenkin chose an academic base at this time underlines his growing reputation as a 'philosophic electrician'. With the science of submarine telegraphy becoming better understood and a consensus developing on best practice, long undersea cables were no longer an experimental technology. While this improved prospects for making money in the industry, for Jenkin it also represented a loss of scope for research. Edinburgh gave him a chance to follow his investigations in new directions, and to consolidate his scheme to place engineering practice on a sound scientific basis.[2]

His new occupation did not confine Jenkin to an ivory tower. It was accepted that a professor of engineering would continue consultancy work alongside university duties. The academic post did confer prestige as well as autonomy, reducing dependence on the whims of telegraph and cable companies, and giving scientific weight to his opinions. Jenkin went to Edinburgh before the furious arguments with Daniel Gooch and Gooch's associates on the French expedition, but it was already clear that a person of Jenkin's temperament, unwilling to suffer fools or temper comments to please his masters, may find academia rather more congenial than the dependant position of a full-time consulting engineer.

---

[1]   Edinburgh University Library, *Senatus minutes*, 29 February 1868.
[2]   W. Thomson, 'Obituary Notices of Fellows Deceased', *Proceedings of the Royal Society*, XXXIX (1885), p. i.

Jenkin's brief attachment to University College London had not been unproductive, and led him towards the Edinburgh post (Figures 5.1 and 5.2). Employed as professor of civil engineering at University College in 1866, he had discovered a talent for teaching and developed a keen interest in education. He reorganised and extended the college's engineering department and was able to persuade the Senate to establish new chairs, in engineering and in applied mathematics and mechanics.[3] Jenkin had also started lecturing on telegraphy

Figure 5.1: Fleeming Jenkin in Edinburgh, *c*.1870
(Jenkin mss)

and electrical measurement at the Royal Engineer Establishment in Chatham.[4]

While abroad on engineering commissions – on road and rail surveys in Saxony, and telegraph works in northern Germany, as well as the Rouen

---

[3]   He was also involved in recommending the influential A.W.B. Kennedy to the UCL chair in 1874. Thanks to Peter Filby for this and other information about Jenkin's connection with University College.

[4]   *Devonshire commission*, p. 94. This connection continued, and Jenkin delivered a paper on telpherage at the School of Military Engineering in the year before his death: Colvin and Ewing (eds) 1887, II, pp. 359, 369.

waterworks scheme – he took the opportunity to examine local educational systems. Once appointed to Edinburgh he embarked on a survey of French, German and Swiss technical education.[5] 'I visited a number of the colleges abroad with the object of ascertaining what was done practically to educate engineers abroad, and to refresh my memory on the subject,' he later told the Devonshire committee.[6] With customary thoroughness, Jenkin observed the training and capabilities of workmen as well as those of professional engineers, visited polytechnics and colleges, looking at syllabuses and methods of training as well as probing the levels of scientific education in elementary and secondary schools. The funding problems at University College, London had more or less obliged him to accept all comers: 'Anybody may come that likes; and I have thus a class of young men of very unequal ability, and I do not think that they profit very much by the lectures'.[7] Fifteen years later, comparing Scottish students with those at University College, London, he recollected that

> the student in London did often wish to get some immediate solution of a particular problem regarding the work with which he was likely to be connected; he just wished to solve that one point. I find here the student is much more patient, and is willing to work from the beginning to the end.[8]

Yet he was said to have done 'such excellent work' at University College, London to recommend him for the appointment at Edinburgh.[9] His experiences there had helped formulate ideas of better ways to organise an engineering education, and the authorities in Edinburgh handed him an opportunity to construct a system closer to these beliefs.[10]

The Edinburgh appointment was to the chair of engineering, specifying that he teach 'engineering both civil and mechanical'.[11] Jenkin, who habitually referred to himself as professor of civil engineering,[12] was still seen as a civil engineer, though occupied for much of the 1860s with electrical problems and systems. Electrical engineering was considered a branch of civil engineering at least until 1871, when the Society of Telegraph Engineers was established.[13]

[5] *Samuelson committee*, pp. 122–42.
[6] *Devonshire commission*, p. 92.
[7] *Samuelson committee*, p. 134.
[8] *Educational Endowments (Scotland) Commission*, Minutes of Evidence, p. 782.
[9] Obituary, *Journal of the Society of Telegraph Engineers*, p. 347.
[10] *Samuelson committee*, pp. 131, 134.
[11] Edinburgh University Library, *Senatus minutes*, 26 June 1868.
[12] *Devonshire commission*, p. 92.
[13] At the foundation of the Society, its first president, William Siemens, felt the need to justify having an organisation separate from the Civil Engineers, referring to the danger of 'degenerating thereby into 'specialists'': *Journal of the Society of Telegraph Engineers*, 1 (1871–2), p. 19.

Jenkin's telegraphic work also contained a strong element of mechanical engineering – designing cable-laying and retrieval systems, for example. In fact the Royal Commission on Technical Instruction, visiting Edinburgh in 1884, noted Jenkin's department as 'mechanical engineering',[14] although by then the professor had come around to thinking of himself as an electrical engineer.[15] He continued to work on wider aspects of civil engineering, particularly on structures, and his contributions to civil and mechanical engineering were arguably as important as those he made to electro-technology. Between 1869 and 1877 he published a number of papers dealing with structures and machines, besides introducing significant new ideas into design methods.

Jenkin's work on bridges shows the scope of his engineering interests at this time. His patents during the 1860s included two for bridge construction, and he later contributed an article on bridges to the *Encyclopaedia Britannica*.[16] Clerk Maxwell had published a theory of reciprocal figures in 1864 which demonstrated a method of calculating stresses and strains in determinate structures, using examples that left 'nothing to be desired by the mathematician'. In a paper read to the Royal Society of Edinburgh in March 1869, Jenkin explained this theory and showed how it could be applied in 'real' situations.[17] More frequently, though, frameworks were indeterminate[18], and different methods were required. Jenkin had come across problems of indeterminate structures in 1861 when pondering the 'correct form' of an iron arch. He touched on the subject of bridges in a letter to William Thomson, in which he sketched an arch bridge constructed from hinged members arranged so that all the forces should be compressive.[19] Jenkin was very pleased with this idea, and patented it the same year. He told Thomson: 'Nothing could be easier than to calculate geometrically the strain on every component part of the bridge, however loaded – turn it upside down and it makes the true form of suspension bridge – as stiff as you please'.[20] His enthusiasm was dampened when it turned

---

[14]   *Royal Commission on Technical Instruction*, p. 493.

[15]   He told the Educational Endowments Commission in 1883: 'I am an electrical engineer myself': p. 787.

[16]   Obituary, *Institution of Civil Engineers Proceedings*, pp. 374–75.

[17]   'On the practical application of reciprocal figures to the calculation of strains on framework', *Transactions of the Royal Society of Edinburgh*, XXV (1869), pp. 441–47.

[18]   A determinate framework is composed of simple triangles. If a girder is so constructed and is supported at its ends and loaded, it is possible to determine the stresses in all the components with the simple methods of statics. Adding an extra diagonal does not necessarily increase the strength of the framework, and may also make the arrangement impossible to analyse by statical methods. This finding emerged from Jenkin's suggestion that a suspension bridge could be treated as an inverted girder bridge. To deal with this, Maxwell devised the method of 'virtual work', calculating the forces by distorting the framework, working out the energy involved and then minimising it. See Jenkin's paper in the Transactions of the Royal Society of Edinburgh for 1869.

[19]   Kelvin, Cambridge, J26, Jenkin to Thomson, 22 February 1861.

[20]   Kelvin, Cambridge, J26, Jenkin to Thomson, 22 February 1861.

out that this method did not in fact work. Jenkin subsequently raised the problem with Clerk Maxwell, who went on to suggest in the *Philosophical Magazine* for May 1864 a method of virtual velocities which positively determined the resultant and all stresses in framed structures.[21]

Jenkin put forward more considered thoughts on bridges to the Royal Scottish Society of Arts in 1869, where he pointed out that girders, arches and suspension bridges were in fact members of the same class.[22] Girders and arches differed only in the observer's perception, while suspension bridges, as he had believed in 1861, were merely reflections of arches. Jenkin thought that Maxwell had supplied the requisite methods: the theory of reciprocal frameworks provided useful graphical techniques, while the use of virtual velocities allowed all frameworks to be subject to calculation. It was later suggested that by advertising the graphic method of treating stresses in a way which made its practical value obvious, and by incorporating it in his teaching, Jenkin had been instrumental in introducing the previously little-known technique to engineers.[23]

The wide range of interests – which also took in heat engines, phonographs, electric motors, control systems and sanitation over the course of his life in Edinburgh – and his ability to marry theoretical with practical aspects of his work, fitted Jenkin well for teaching a broad-based engineering course. In his evidence to the Samuelson Committee on Scientific Instruction, Jenkin described himself as having been both a civil and a mechanical engineer. 'I began life as a mechanical engineer, as a pupil of Mr William Fairbairn ... and afterwards practised as a civil engineer'.[24] To the Galton committee in 1860, Jenkin introduced himself as a civil engineer, unlike Cromwell Fleetwood Varley who already called himself an electrical engineer.[25] Jenkin had first entered the Institution of Civil Engineers in 1859, becoming a full member in 1868. His proposer for 'the Civils' was William Fairbairn, and supporters included another former employer George Hemans; Brunel's associate on the building of the *Great Eastern* John Scott Russell; and two others connected with submarine telegraphy, Lionel Gisborne and William Siemens.[26] When Jenkin eventually joined the Institution of Mechanical Engineers in 1875, his sponsors again included Siemens, and another old employer, John Penn. Both

---

[21] Jenkin refers to this in his paper, 'On braced arches and suspension bridges', *Transactions of the Royal Scottish Society of Arts*, VIII (1870), p. 138.

[22] This was published as 'On braced arches and suspension bridges' in their *Transactions* the following year.

[23] Obituary, *Institution of Civil Engineers Proceedings*, p. 370.

[24] *Samuelson committee*, p. 122.

[25] *Galton committee*, pp. 135, 148. For details of professional engineering institutions in the nineteenth century, see R.A. Buchanan, 'Institutional Proliferation in the British Engineering Profession, 1847–1914', *Economic History Review*, XXXVIII (1985), pp. 42–60.

[26] Institution of Civil Engineers, membership form A, 29 November 1859.

Fairbairn and Penn had been involved on the periphery of submarine telegraphy, Fairbairn as a Board of Trade appointee to the Galton committee in 1859. The firm of John Penn, a long-time advisor to the Atlantic Telegraph Company, had built and set up the cable machinery on the *Great Eastern*.[27] Jenkin maintained contact with these and other eminent practitioners across the whole spectrum of engineering, before and after his move into academic life in 1868.

The chair of engineering at Edinburgh was a newly established post in 1868, although the university had earlier had a Regius Professor of Technology. George Wilson (1818–59) had been appointed in 1855 to that chair, in combination with directing the Industrial Museum of Scotland. Wilson, after training as a surgeon, had practised and taught chemistry, as well as presenting science to the public through lectures and articles in popular quarterlies.[28] Although his courses in the university attracted large numbers of students, the chair was abolished on his death.[29] In a eulogy to Wilson, praising him as an able teacher, Sir David Brewster (1781–1868), vice-chancellor at Edinburgh from 1859 and the first physicist to head a British university, suggested in 1860 that the university should embrace the teaching of civil engineering in all its most modern manifestations, emphasising the importance of mathematics and science as essential requisites for the engineer.[30] Brewster, instrumental in establishing the Regius Chair in 1868, died days before Jenkin was appointed to Edinburgh.[31] Some decades earlier, Brewster had expressed the view that 'there is no profession so incompatible with original inquiry as a Scotch professorship, where one's income depends on the number of pupils'[32] and he took care that the chair which Jenkin was to occupy was adequately funded. Sir David Baxter, a Dundee textile manufacturer, had endowed it with £5000, to which the Treasury agreed to contribute £200 *per annum*. The first professor was to be Baxter's nominee, subject to Crown approval. Baxter told the university Senate in February 1868 that he

> hoped it would be made obligatory on the new Professor to teach as part
> of the course the application of Engineering to manufacturing industry. It

[27]   Bright 1974 [1898], pp. 41, 59, 86–87.
[28]   See Marinell Ash, 'New Frontiers: George and Daniel Wilson', in J. Calder (ed.), *The Enterprising Scot: Scottish Adventure and Achievement*, (Edinburgh: HMSO, 1986), pp. 40–51. For this reference thanks to Alison Morrison-Low.
[29]   This has been attributed to political causes connected with the museum rather than the university: Morrell 1976, p. 75.
[30]   Morrell 1976, p. 61.
[31]   Brewster also established the first BSc degree in Scotland at Edinburgh in 1864: see Robert Anderson, 'Brewster and the reform of the Scottish universities', in A.D. Morrison-Low and J.R.R. Christie (eds) 1984, p. 33.
[32]   Quoted by Steven Shapin, 'Brewster and the Edinburgh career in science', in A.D. Morrison-Low and J.R.R. Christie (eds) 1984, p. 17.

would also be pleasing to him if the Professor would make practical excursions with his students, taking them into the field to teach them surveying and going with them to large workshops and manufactories for the purpose of bringing them into direct contact with industrial operations.[33]

Professor Lyon Playfair, writing to Baxter on behalf of the Senate when the chair was first proposed, had described it as 'the means probably of founding a division of industrial science in the University and of giving a great impulse to technical education'.[34] It has since been seen as a significant element in the new movement to establish technical education.[35]

Jenkin's appointment was made by a committee consisting of six of Edinburgh's thirty four professors – John Wilson, Robert Christison, P.G. Tait, Alexander Campbell Fraser, William Young Sellar and Lyon Playfair – and attended by Sir David Baxter.[36] There were three applicants, the other leading candidate being W.J. McQuorn Rankine, Professor of Civil Engineering and Mechanics at the University of Glasgow since 1855.[37] While Jenkin had much shorter experience as an academic than Rankine, his scholarly credentials were more than respectable and his work had novel and practical qualities which presumably appealed to the Dundee industrialist. His Edinburgh background and top-level engineering connections may have helped, and he had no difficulty in accommodating Baxter's advice on teaching methods. Jenkin had a solid workshop education and was used to practical tuition, from his surveying lessons in Genoa and his experiences in Bancalari's laboratory. The six professors recommended Jenkin, and Baxter concurred.

Baxter's endowment had been split into £1000, yielding forty pounds a year for class expenses, and £4000 towards the professor's salary. The £200 *per annum* which this raised was matched by the Treasury.[38] In addition there were class fees of about £200. While this was considerably less than the £1000 a year which Gladstone believed professors should receive,[39] Jenkin's income from the

---

[33] Edinburgh University Library, *Senatus minutes*, 29 February 1868. The same wording was incorporated into the Deed which formally founded the chair and appointed Jenkin to it: idem, 26 June 1868.
[34] Edinburgh University Library, *Senatus minutes*, 20 January 1868.
[35] Morrell 1976, p. 76. See also Sanderson 1972, p. 171, regarding the Baxters' contributions to education in Dundee.
[36] Edinburgh University Library, *Senatus minutes*, 29 February 1868. The committee had met on 27 February. The third candidate was Professor Coote.
[37] Edinburgh University Library, *Senatus minutes*, 29 February 1868. While Rankine (b.1820) had been professor of engineering in the University of Glasgow since Gordon's resignation in 1855 and was renowned as 'a great engineering educator', he had also been a practising engineer since 1838: see Emmerson 1973, pp. 120–21, and Sanderson 1972, p. 160.
[38] Edinburgh University Library, *Senatus minutes*, 25 January 1868.
[39] Birse 1983, p. 102.

university was far higher than that of some of his colleagues. His contemporary at the Edinburgh Academy, Peter Guthrie Tait, who had been appointed to the long-established chair of natural philosophy in Edinburgh in 1860, did not receive a salary and his reliance upon student fees compelled him to admit all applicants. Jenkin used the example of Tait – 'an admirable professor' forced to take 'wretchedly prepared students' – in his evidence to the Devonshire committee in 1870.

> Now a man like Professor Tait is thrown away teaching elementary science, which ought to have been acquired in the second grade schools, and if we are ever to have a much higher grade of teaching in the University, we must have entrance examinations, and you cannot make these men insist upon entrance examinations if they are to do so at the penalty of losing three-fourths of their income.[40]

Moreover, professors did not receive the whole of their salaries and fees, for they were expected to meet personally the costs of equipment and assistants.

Low pay for academics was a wider problem. The prestige of the job was not matched by material rewards.[41] Sir David Brewster had been paid £700 *per annum* as Principal of Edinburgh University, which the Senate considered 'utterly inadequate' to attract a suitable replacement after his death.[42] In the mid-nineteenth century, fellows at Oxford or Cambridge received a dividend of only £200 a year.[43] A scholar as well-known as T.H. Huxley was obliged to collect a number of academic and museums posts to enable him to marry: 'A man who chooses a life of science chooses not a life of poverty, but so far as I can see, a life of nothing, and the art of living upon nothing at all has yet to be discovered.'[44] There was, though, more money to be made teaching engineering. One of Jenkin's students, Alfred Ewing, who claimed not to be 'a heaven-sent engineer', was appointed in 1878 at the age of twenty-three to a chair of engineering in Tokyo which paid 370 silver dollars – about seventy pounds – a month, plus a house, travel expenses and a good laboratory.[45]

A further advantage of engineering over pure science was the possibility of industrial consultancy work. In the words of Brock, this provided compensation for Jenkin 'for the lack of patronage towards engineering shown by the University of Edinburgh'.[46] It is implied that his income could have been

---

[40]   *Devonshire commission*, p. 98.
[41]   Newsome 1997, pp. 70–71.
[42]   Edinburgh University Library, *University Court Minute Book*, I, 3 April 1868.
[43]   By comparison, senior masters at good public schools earned much more, from about £900 up to £4000 for the head of Eton during the closing decades of the century: Newsome 1997, p. 70.
[44]   Newsome 1997, pp. 70–71.
[45]   Kelvin, Cambridge, E106, Ewingto Tait, 30 November 1882; Ewing 1939, ch. 5.
[46]   Brock 1976, p. 187.

much higher had he concentrated upon commercial work alone. Yet the university did provide him with a base as well as a modest assured income, and also gave the opportunity to meet some of the brightest and best fledgling engineers of a new generation. This last, for a man forever young at heart and intellectually curious, may have been the greatest inducement of all to locate himself in academic life.

As Jenkin's only contractual obligation was to teach during the winter session, which ran from November to April,[47] the northern hemisphere's summer cable-laying season was conveniently free for consultancy work. Practical experience on telegraph voyages could not be offered to all undergraduates, but Jenkin was able to employ his best students as assistants, as he had with Leslie Hill from University College, London on the French Atlantic expedition in 1869. This provided income as well as training for undergraduates, some of whom were in financial difficulties. It had been noted by an enquiry in 1858 that 'a very large number of the students in the Scotch Universities are in exceedingly poor circumstances' and that many had to teach or take other employment in summer to support themselves at university through the winter.[48] Alfred Ewing, (1855–1935), although from an educated background as a son of the manse, was able to go up to Edinburgh in 1871 only through a Baxter Scholarship from Dundee High School.[49] In the summers of 1872 and 1873, Jenkin employed Ewing on electrical testing in Hooper's cable factory in London, at first paying only expenses, though later he received a small salary.[50]

Despite the improvements in Jenkin's own personal and professional circumstances, in the university money and space remained in short supply. A number of new professors had been appointed in the sciences, and for engineering there had been no funding to provide basic and essential facilities. In the older chair of natural philosophy, Tait, who was interested in physical rather than purely mathematical problems, managed to achieve a reputation for laboratory work 'of a rarely equalled magnitude and importance'.[51] Yet during his first eight years as professor, Tait did not have a laboratory, and was obliged to conduct experiments in his classroom and private room in college. While this was in the tradition of his predecessors, some other institutions had already acquired laboratories: William Thomson's in Glasgow dated from about 1850.[52]

[47]   Edinburgh University Library, *Senatus minutes*, 26 June 1868.

[48]   *Devonshire commission, final report* (1875), p. 331. See Sanderson 1972, pp. 151–4.

[49]   These grants of £40 a year to two students for up to two years each, in the gift of the High School's directors, had been endowed in March 1869 by Miss Baxter, apparently at the instigation of Jenkin and her brother: Edinburgh University Library, University Court Minute Book, I, 9 August 1870.

[50]   Ewing 1933, p. 258.

[51]   *Dictionary of National Biography*.

[52]   Knott 1911, p. 22.

Tait's laboratory was part of a new wave of such facilities between 1866 and 1874, when ten were established in British universities; the following decade saw a further fourteen new physics laboratories.[53]

Tait had managed to secure a small grant from university funds in 1867, and his Physical Laboratory opened the following year. This grand title hardly matched the reality. Tait's celebrated laboratory was in fact a converted pathology class-room housed in the university's eighteenth-century building. He described it as:

> a small classroom which had come to be disused, entirely unsuitable, or at least by no means very suitable for almost any class of experiments.' The result of this is that when more than 8 or 10 students attend the Laboratory at once, some of them are obliged to work in the Class-room, and some among the Professor's collection of apparatus. 'The superintendence of groups of students scattered about, with stairs to ascend, and passages between them, is a matter of considerable difficulty, and adds materially to the labour of teaching.[54]

Conditions were so cramped that it was impossible to use the powerful electromagnet without disturbing galvanometers, and Tait's instruments were piled 'in strata' around the room.[55] In other practical classes, students were turned away from courses because there was not sufficient space for them – a serious matter when part of the professor's income came from student fees. Tait also complained that he lacked assistants, both technical and teaching. In 1872 he had one mechanical assistant, a man in his seventies who had served as doorkeeper at the university; and a class assistant who helped with tutorials and marking, as well as with supervising the laboratory.[56] The shortage of support meant that Tait was forced to spend nine months of the year on teaching, leaving only about three months for research and holiday.[57]

While Tait's facilities were highly unsatisfactory, Jenkin's were worse, for the engineering department had no laboratory at all. Jenkin was forced to improvise, though the resultant method of teaching probably suited him besides complying with the wishes of Baxter. Jenkin devised courses which could be carried out in converted classrooms, such as mechanical drawing, or outside the university, like practical surveying.[58] This was a man, after all, whose early

---

[53]   Smith and Wise 1989, p. 653. See also Gooday 1991.
[54]   *Devonshire commission, final report*, p. 337. Tait had given his evidence to Devonshire on 16 February 1872.
[55]   Morrell 1976, pp. 79–80.
[56]   *Devonshire commission, final report*, p. 339.
[57]   Morrell 1976, p. 81.
[58]   Morrell 1976, p. 80.

experiments in electricity had been carried out in a factory in Birkenhead under pressure of production deadlines.

In Edinburgh there was no special accommodation for engineering, and the new department was allocated limited lecture-room and drawing office space in a top corner of the Old College. It was to be more than twenty years before an engineering laboratory was equipped. But Jenkin had the advantage over Tait of an income as a consulting engineer which enabled him to employ numbers of assistants and continue his research, much of which took place at his home.[59]

Tait was a long-standing acquaintance, and Jenkin openly respected his great intellect, yet relations were strained. Between the two was a philosophical and temperamental divide. Tait objected to the wide range of Jenkin's interests, seeing his own more theoretical work as superior. At the same time he envied Jenkin the high income derived from consultancy work, which paid for teaching relief and enabled Jenkin to travel away from Edinburgh on other business.[60] He also saw Jenkin as an obstacle to his own collaborative work with Sir William Thomson, for Jenkin and Thomson's lucrative consultancy business took up much of Thomson's time.[61] As a young man Alfred Ewing observed the 'sharp competition' between Tait and Jenkin for Thomson:

> [Thomson] was Jenkin's partner in practical concerns that involved big responsibilities and clamoured for attention. He was Tait's partner in the authorship of Thomson and Tait's *Natural Philosophy* – a gigantic infant that seemed always struggling to the birth. Hence between Jenkin and Tait there was strife for Thomson's soul. In point of fact Jenkin did all that could be done to relieve his partner of business detail. But in Tait's eyes Jenkin stood for a malign influence dragging Thomson to earth when he should have been free to soar and float in the serene air of mathematics, at a level where Natural Philosophy might forget that it had anything to do with the affairs of men.[62]

To pile on further insult, Jenkin and Thomson carried out some of their telegraphic experiments in the very laboratory which Tait had spent years

---

[59] Ewing 1939, p. 35; Morrell 1976, p. 84. Problems of laboratories did not recede: when Ewing was appointed to the chair of engineering at the new Dundee College in 1883, Jenkin wrote to him 'pooh-poohing the idea of starting an engineering laboratory at first': Ewing 1939, p. 74. Ewing went to Cambridge in 1891 as Professor of Mechanism and Applied Mechanics, to find that there were workshops but no suitable lecture room or laboratory – in his inaugural lecture Ewing set out the acquisition of a laboratory as one of his priorities: Ewing 1939, pp. 98–9.

[60] Morrell 1976, p. 81.

[61] For details of Tait and Thomson's various collaborations, see Smith and Wise 1989, especially Chapter 11 on their *Treatise on Natural Philosophy*.

[62] Ewing 1933, p. 259.

scraping together money to equip. Jenkin's view was that Tait, with his narrow scientific pursuits, 'voluntarily abjures some of the best parts of life'.[63] Being Jenkin, he would not have kept his opinion to himself, and Tait's resentment is easy to understand. Tait too was outspoken; Robert Louis Stevenson considered that he had 'no tact and no manners'.[64]

Although Jenkin spent much time on consultancy and other interests, he did not neglect his professorial duties nor forget his concern to advance technical education. In fact educational interests increasingly absorbed him from the mid-1860s. 'Technical education', it was said, 'much interested him long before it acquired its present interest for the public', evidenced by his attendance at the Society of Arts and other bodies when the subject was discussed.[65] The Society of Arts had organised a conference on technical development in January 1868, attended by Playfair, Huxley, Rankine, Samuelson and other luminaries, from which a sub-committee was appointed to consider various proposals. Jenkin found himself elected to this group, which met twenty-six times in the following months. Their report, published in July 1868, defined technical education as 'general instruction in those sciences, the principles of which are applicable to various employments of life' while specifically excluding 'the manual instruction in Arts and Manufactures which is given in the workshop'.[66] The committee's recommendations included a more systematic approach to science education – the introduction of new science schools and public examinations. Many of their proposals were directed towards improving part-time educational facilities for shop-floor workers, those skilled artisans who had so impressed Jenkin during his time in Birkenhead.[67]

In Edinburgh, Jenkin became a staunch supporter of the Watt Institution and School of Arts. This, the forerunner of Heriot-Watt University, founded in 1821 as the Edinburgh School of Arts, is acknowledged as the first mechanics institute in Britain.[68] Although it did not bear that name, its purpose was exclusively the education of the working class, and it was funded by public subscription and class fees. Its early curriculum was deliberately confined to subjects of direct use to mechanics and other workmen – Chemistry, Natural Philosophy, Mathematics – but this was gradually expanded over the years so that modern and ancient languages and other sciences were offered by the

---

[63]   National Library of Scotland, ms 2633, ff. 208–10, Jenkin to William Young Sellar, 21 May 1880.

[64]   Booth and Mehew (eds) 1995, VI, p. 140.

[65]   Obituary in Nature, XXXII, 18 June 1885, p. 154.

[66]   Journal of the Society of Arts, 24 July 1868, XVI, p. 627.

[67]   Journal of the Society of Arts, 24 July 1868, XVI, pp. 627–33. See also Emmerson 1973, p. 176.

[68]   See Heriot-Watt University, 1973, pp. 1–18. Leonard Horner, instrumental in founding the School of Arts in 1821, was also a founder of the Edinburgh Academy in 1824 and became first Principal of University College, London, in 1826: ibid.

1870s. A report on the school in 1877 notes that 1400 students had enrolled in the previous session, mainly 'young men in their apprenticeship, or who have just passed that stage'.[69] The fees were five shillings per class for the session. In 1873, when Fleeming Jenkin was co-opted as a director, the institution had recently moved to impressive new premises in Chambers Street. Although Jenkin often missed directors' meetings during the 1870s, he used his experience to help in practical ways.[70] As the Watt Institution had no principal, directors took on a wide responsibility.[71] Jenkin shortlisted and interviewed prospective drawing masters, checked accommodation and teaching standards in a number of classes, and solicited books for the library from publishers.[72] From 1880 he became involved in the campaign to persuade the Scottish Educational Endowments Commission to change the status of the institution to that of a technical college.

Jenkin's admiration for the Watt Institution was recorded in his evidence to the Devonshire Commission as early as 1870. This 'admirable little institution' ran on £390 a year, of which only £190 was donated, the rest being students' fees.

> The results that it produces from that small expenditure are marvellous. I have had several students from there; they are all well trained in mathematics, and have a very fair knowledge of the elements of natural philosophy.[73]

This despite the 'small room and miserably scanty apparatus with which such very good results have been obtained'. Attending the Educational Endowments enquiry in 1883 as representative of the Watt Institution's governors, Jenkin made a plea for better funding so that classes could be reduced in size and streamed, and new subjects offered, especially sciences, with better facilities. Yet he did not want radical changes, arguing that the Watt should build upon its traditions and continue with the same type of student. He saw its working class base as a strength. 'If we change the character of the place very materially we may lose the people we most want to get.'[74] He feared that if daytime classes were substituted for evening teaching, then

---

[69] Heriot-Watt University Archive, SA1/2, *Minute Book of Directors*, November 1877.
[70] Jenkin was also a director of his old school, the Edinburgh Academy, from 1874 to 1885, where his attendances at meetings were similarly sporadic and his main contribution was to help with practical matters such as arranging the teaching of engineering drawing and advising on heating systems and other building works. For searching through *the Academy Directors' Minute Books* on our behalf, we are grateful to Brian Cook.
[71] Educational Endowments (Scotland) Commission, p. 789.
[72] Heriot-Watt University Archive, SA1/2, *Minute Book of Directors*, 2 August 1875; 4 May 1877; 15 November 1877; January 1878; April 1878.
[73] *Devonshire commission*, pp. 96–97.
[74] *Educational Endowments (Scotland) Commission*, p. 783.

the technical school would become another college for the upper classes. I don't think it would teach workmen. It is desirable that the workman should be at his work during the day and that he should get theoretical instruction in the evening ... I think that the instruction intended for workmen and foremen should be of a different kind from the scientific instruction for the professional classes, and that the two should not clash.[75]

While the teaching at the Watt Institution should not overlap that of the university, Jenkin approved of successful students moving on to degree courses. Artisans motivated to attend evening courses he saw as the very best of their class: 'Their very attendance upon those classes proves that they are remarkable men'.[76]

It not infrequently happens that the best students from the Watt Institute come up to the university, and I have frequently had students taking prizes in my class who previously took prizes in the Watt Institute.[77]

He was subsequently appointed one of seven Life Directors to run the newly constituted Heriot-Watt College from 1885. The part which he had played in bringing about this change was acknowledged in a tribute by the chairman of the new board of Heriot-Watt immediately after Jenkin's death; Lord Shand lamented the college's great loss, saying that he had looked to Jenkin 'as the principal man to assist in organising the Institution under the new scheme'.[78]

Confirming the importance he attached to the subject, Jenkin had chosen to speak on the education of civil and mechanical engineers in his inaugural lecture in Edinburgh in November 1868, when he advocated a middle course in which apprenticeship should follow university, instead of the conventional engineering training which was entirely industrial.[79] He gave a paper on technical education to the Royal Scottish Society of Arts the following year, and in 1871 selected the same topic for his presidential address to the Mechanical Section of the British Association. He explained his educational philosophy to two parliamentary enquiries, the 1868 Select Committee on Scientific Instruction chaired by Bernhard Samuelson, and the long-running Royal Commission on Scientific Instruction and the Advancement of Science under the chairmanship of the Duke of Devonshire, before which Jenkin appeared in 1870. Jenkin was firmly opposed to college education being used as

---

[75]　*Educational Endowments (Scotland) Commission*, p. 782.
[76]　*Devonshire commission*, p. 101.
[77]　*Educational Endowments (Scotland) Commission*, p. 781.
[78]　Heriot-Watt University Archive, SA1/2, *Minute Book of Directors*, 24 July 1885.
[79]　Sanderson 1972, p. 14.

a substitute for apprenticeship or pupilage.[80] It is not, however, true to say that his educational stance was conservative; nor is it right to place him in a 'traditionalist alliance' of early telegraphers on the subject of training engineers.[81] Jenkin's opinions on education were entirely his own, and were informed by wide investigation of British and continental systems. His inaugural lecture has been praised as a 'wise and clear statement [which] set the pattern of the British approach to higher engineering training to this day'.[82]

His scheme to improve engineers' education and training was founded in two considerations. First, he took an overview which encompassed all classes of engineering; his perspective on education, like his ideas on telegraphy, was all-embracing. His views about the education and training of working men were not an add-on but part of a wide-ranging concept of industrial training which extended to include Tait and his ilk. The second consideration which underpinned Jenkin's educational philosophy was that he was a political realist and ready to acknowledge which parts of the system would not or could not quickly change. He saw, for instance, that British industry would not tolerate a plan such as that operating on parts of the continent, where professional engineers did not qualify until the age of twenty-five. Better informed than most about continental methods, Jenkin did not swallow the idea that they were wholly superior to those in Britain. Moreover, he knew the restrictions of time, space and money in British universities and acknowledged that a new scheme of engineering education could not be rapidly superimposed upon the older institutions. Jenkin's proposals, while informed and critical, were never unrealistic.

Jenkin's thoughts on education took in the whole range of practical and theoretical training, at every level in both professional and artisanal engineering. He suggested that standards and entry requirements at each stage needed improvement – ridding the system of Tait's 'wretchedly prepared' students. Once he was himself convinced, he worked hard to persuade others to his philosophy.

> Fleeming Jenkin's work as an educator was far from being bounded by the walls of his classroom ... He contended with much vigour and force that the system of apprenticeship should be maintained, that attempts to substitute for it the practice possible in a merely educational workshop were futile; but that it required, as preparation, such training in mechanical science as could and should be given in the common universities and schools. He ... urged that the Universities should offer a

---

[80] *Devonshire commission*, p. 93.
[81] See Gooday 1991, especially pp. 81 and 105, for the suggestion that Jenkin was allied with Rankine and Whitehouse in a rearguard action to resist teaching telegraphy in laboratories.
[82] Sanderson 1972, p. 14.

complete theoretical course and grant degrees in engineering; that on the other hand engineers should look for a fair knowledge of theory in the pupils they took, and give a preference to those who came prepared. He insisted that mechanical drawing should be taught in primary schools, as a first requisite in the education of every working man.[83]

Jenkin saw the failure to teach mechanical drawing as a fundamental fault in the British education system, compared with continental practices where the subject was available to working men in almost every town and village. He thought that its significance was not understood.

> It is almost as important to a skilled workman as reading or writing or arithmetic. If you understand mechanical drawing you can read all the books which are written about machinery, and understand them if they are not deeply mathematical; you can read all the engineering newspapers which contain a vast amount of information, and you can explain your own ideas, and put them down, and the mere power of recording their ideas gives the workmen a very much greater power of thinking.[84]

Jenkin had initiated a course in mechanical draughting at University College, London, taught by a more junior tutor under his supervision.[85] A similar programme in mechanical drawing was among his innovations in Edinburgh .

Jenkin's model for training professional engineers was a proto-sandwich course which resembled those at the most modern continental polytechnics. His ideal system would have well-prepared students first attending pure science courses, followed by a pupilage, at the end of which they would spend six months attending specialised technical courses such as those which Jenkin taught – the application of theoretical knowledge to practical demands.[86] The first part of this training, a theoretical grounding – 'the mere mother sciences, as I call them – mathematics, chemistry and natural philosophy, chiefly' – would replace the existing degree taken usually between the ages of sixteen and eighteen.[87] A scientific foundation common to all branches of engineering would eliminate the woeful ignorance which Jenkin had found even in that most scientific branch of engineering, telegraphy. Responding to questioning by the Samuelson Committee about the amount of scientific knowledge required in the higher ranks of engineering, Jenkin claimed that there had been eminent British engineers who possessed no knowledge of science:

---

[83] Obituary, *Institution of Civil Engineers Proceedings*, p. 373.
[84] *Devonshire commission*, p. 95.
[85] UCL *Annual Report*, 26 February 1868, pp. 10–11.
[86] *Devonshire commission*, p. 100.
[87] *Devonshire commission*, p. 93.

There have been eminent engineers guiltless of algebra ... We have had
such men who could not calculate the strain on any portion of a machine;
they have acquired mechanics as an art, not as a science.[88]

This was no longer tolerable, said Jenkin. A young engineer who lacked
scientific education could not progress to greater responsibility.

I think that in the future a man without theoretical knowledge will be at a
considerable disadvantage; in the early days of a profession a man of
sound common sense only will rise to the top, but the engineering
profession is becoming more and more complex.[89]

In his letter of advice to the young Forbes in 1865, Jenkin had
differentiated between two possible paths in professional engineering. For the
less innovative kind of engineer, he suggested, a commercial and linguistic
education could prove more useful than advanced technical training:

If a Civil Engineer means simply to follow the beaten track, without any
expectation of rising to an independent position by his own ability,
Mechanical Engineering will be of little use to him. Contractors do the
designing nowadays with the aid of draughtsmen at two or three guineas
a week and of a few Engineers who are seldom gentlemen but generally
self made men of considerable ability. Engineers duties are more of an
Inspector and man of business. As a young man he must see that
contracts are properly carried out and as an old man that they are
properly drawn up and to do this he must have some experience as to the
practices of contractors with a knowledge of arithmetic simple equations
a few problems of plane trigonometry and the use of logarithms; so that
he may understand how to apply the rules for calculating quantities etc
which he will find printed in his Engineers pocketbook. He will find it
advantageous to have a knowledge of languages, to have a liberal
education (especially if he ever rise to meeting those august bodies
Boards of Directors face to face) and his employers will be sure to find
him useful if he is methodical, understands how to write and construe
English with accuracy i.e. knows what he means to say and says it; and
understands what other people say and write. You will see that in this
sketch there is little of the ingenious contriving inventive mathematical
animal which novels depict. Mechanical Engineering as gained in a
Workshop is of no use to this class of man. He must go into an Engineers
office in *London*; I cannot say which is the best with a premium paid and
with a special understanding that he is to be employed on Works and not
in the Office. If he is a useful man in four or five years he may hope for

[88] *Samuelson committee*, pp. 138–39.
[89] *Devonshire commission*, p. 93; *Samuelson committee*, p. 126.

an appointment as Resident Engineer in charge of some works at £100 or £150 per annum. He will gradually rise to £500 per annum or a little more in England and perhaps double that abroad. Beyond that I cannot follow him.[90]

Electrical engineering, though, as a new and scientifically based branch of the profession, needed more of the 'ingenious contriving inventive mathematical animal' with a sound theoretical understanding. Later, in pleading for more specialist technical courses, Jenkin predicted a severe shortage of electrical engineers. 'The demand will be very great indeed – very great ...' He foresaw new kinds of electrical engineering:

Young men ... will be wanted both to erect electrical apparatus and to superintend its working – the same sort of employment will grow up in respect of electricity as at present exists in respect of steam ... There will be first of all lighting, which will be very considerable; and then the distribution of power. It is quite within a measurable distance that for a large number of purposes the distribution of power by electricity will come more and more into use.[91]

The second element in his scheme was a more methodical period of pupilage. Under the existing system, a youth became a pupil engineer at seventeen or eighteen, having paid a premium, 'that being the sole condition of entering the workshop of a mechanical engineer', Jenkin told the Samuelson enquiry. These 'wonderfully unprepared' boys knew little mathematics and most could not use the formulae in the *Engineer's Pocket Book*. There were few facilities to teach them their profession.

In the first place, no one teaches them anything: neither the master nor the foreman. The foreman thinks it his duty to get as much work as he can out of those who will work, and consequently keeps them as much as possible to one kind of work, however uninteresting it may be.[92]

Jenkin recollected his own time at Fairbairn's, where he had passed six weeks doing nothing but polish brass valve boxes for locomotives. In such a system, he said, many would learn nothing – 'I should say that seventy five per cent of them waste their time'[93] – although the more energetic and able pupils could acquire a practical knowledge of machinery which was not available in

---

[90] University of St Andrews Library, *Forbes mss*, correspondence (in) 1865/62, Jenkin to Thomson, 5 July 1865. Sanderson read this letter as suggesting that Jenkin thought university education unnecessary for an engineer: 1972, p. 13.

[91] *Educational Endowments (Scotland) Commission*, p. 787.

[92] *Samuelson committee*, p. 123.

[93] *Samuelson committee*, p. 137.

any other way: 'With the best men the result is better than you could possibly expect'.[94] Civil engineers – among whom he presumably numbered telegraph engineers – received rather better training than mechanicals as there were useful minor tasks to do alongside the experienced professional.

The third stage of his ideal system was a return to university, after pupilage or during the final six months, to take courses in applied engineering. Jenkin had been impressed by the commitment and achievements of part-time students who were already working in engineering:

> I have men in Scotland, who are very hard-working, who come to me and attend perhaps one other course in the university and do their professional work at the same time, but their habits are so remarkable as to hard work, that I do not know whether anywhere else it could be done. I have men coming and attending my course at 9 o'clock in the morning, going to their work at 10, going on with their work, with only the dinner and tea hour, till 9 o'clock at night, and then doing my exercises and keeping near the head of the class ... they are not workmen, they are men who are in offices as draughtsmen, and as assistant engineers. In the borough engineer's office I have one case.[95]

This class of student had not existed at University College, London – 'my students did not work there'. Jenkin tried to introduce the post-pupilage specialist training at Edinburgh, but had found only one company which would insist upon this extra course as a condition of apprenticeship. The requirement was that the trainee attend two winter sessions during the five years of articles. 'That, I think, will work exceedingly well. I would rather have them at the end, but in the middle is better than not at all.'[96]

The three-stage model of training drew upon Jenkin's experience of teaching at University College and the contrast in Edinburgh, where he saw the value of both a scientific foundation and experience of practical engineering in order to derive the most from his kind of teaching.

> Mine is a technical course. I should like them to take pure science first, and then come to the technical teaching ... a man ought not to come to me before he has been at real work. The students who, I find, practically benefit from what I teach them are men who have already been in contact with real work.[97]

---

[94] *Samuelson committee*, p. 123.

[95] *Devonshire commission*, p. 100.

[96] Jenkin here refers to 'apprenticeship' and it is not clear whether he means pupilage: Devonshire commission, p. 100.

[97] *Devonshire commission*, p. 100.

This final element of his scheme, the technical education to be taken at the end, or immediately after completion, of a pupilage, demanded the appointment of more professors. Jenkin preferred that any extra spending on professors should go to new chairs and lectureships, rather than raise the salaries of existing incumbents, so that chairs could be endowed in the different branches of engineering and each would be seen as a speciality.[98] Ideally, the foundation element would be carried out in schools, with universities delivering the special courses after pupilage.[99] He emphasised that he did not wish to set up technical schools, within or outside universities, but believed that some technical courses could usefully be added to 'the purely scientific teaching now given in universities'.[100] He strongly resisted the notion that engineering should be taught in separate institutions, seeing advantages in a proximity to other

Figure 5.2: Fleeming Jenkin in Edinburgh, c.1880
(Jenkin mss)

studies, whether in a university or at the Watt Institution.

I would rather see my own class form part of an institution which is partly literary; and I would rather see in the people's college a literary

---

[98] *Devonshire commission*, p. 94.
[99] *Devonshire commission*, p. 103.
[100] *Devonshire commission*, p. 105.

side not separate from science. I would rather carry it on as a university.[101]

This is consistent with an opinion expressed to the Devonshire Commission, where Jenkin had suggested that his chair was 'the better for being in the university. I think my students are better for the students that they meet. I think that I myself am better for the professors that I meet with and associate with'.[102]

Jenkin's model of engineering education was not one which flatly opposed practical training in universities. He admired the fellow pupil at Fairbairn's who had received a scientific education as well as some workshop instruction at King's College.[103] His objection was to the system practised on the continent, where there was no pupilage and all engineering training took place in colleges.[104] This opinion was based upon pragmatics, as Jenkin judged that the British system, with all its shortcomings, still produced better engineers than even the best German polytechnics. He was convinced that industrial conditions could not at that time be usefully replicated in the laboratory. He was not on principle attempting to divide the teaching of theory from practical work. Indeed Jenkin concurred with Thomson and his Glasgow circle in emphasising a harmony of theory and practice in the making of a professional engineer.[105] He did not see theory and practice as opposing each other, or as in competition for superiority, or for that matter as overlapping – they were separate things. 'But what is the use of theory if you cannot put it into practice?'[106]

How much of this educational model was Jenkin actually able to translate into practice? In Edinburgh, he started with the advantage of launching a new degree course with a good deal of autonomy. He had Baxter's backing and believed that Scottish students would come to him better prepared than had been the case in London. He had already rejected some applicants, and had set up 'a proper matriculation examination, such as would be required for medical men ...'[107] His two year course, for undergraduates aged from sixteen to eighteen, aimed to be highly theoretical and would emulate continental preparatory courses.

If we could get students to enter upon their pupilage as well prepared as [continental students] are when they enter on those special courses, I

[101] *Educational Endowments (Scotland) Commission*, p. 788.
[102] *Devonshire commission*, p. 103.
[103] *Samuelson committee*, p. 122.
[104] *Devonshire commission*, p. 93; *Samuelson committee*, p. 127.
[105] Smith and Wise 1989, p. 654.
[106] *Devonshire commission*, p. 95.
[107] *Samuelson committee*, pp. 134, 140.

think we should then get a class of very much better engineers; a class possessing, in fact, all the theoretical knowledge we need.[108]

Jenkin went on to specify what he intended to include in this foundation course:

> You want to teach them mathematics as far as they are required to enable men to apply them to ordinary engineering problems ... You cannot take the whole range of engineering, but you can show them, for instance, how geometry is applied in mechanical drawing. In my own course I can show them how physics and chemistry, and mathematics, are all applied in certain definite examples. I may take bridges and locomotives, and show the application of science to these examples somewhat fully; and then, if a student has common sense, he will see how those different elements are applied in practice to other examples ...[109]

The University of Edinburgh was already offering a postgraduate diploma, gained by examination after pupilage, which Jenkin hoped would become a more widespread test of proficiency for newly qualified engineers.[110]

The course syllabus of the mid-1870s shows that Jenkin had tried to apply his model as far as was practical.[111] Students took their Engineering degree in two stages. First was the foundation in theoretical science, where examinations were given in mathematics, natural philosophy – consisting of applied maths and experimental physics – and chemistry. It was possible to gain exemption from the foundation and move straight to the second stage, where applied maths, engineering and drawing were studied. The engineering course drew upon the main branches of the subject, including properties of materials; strength, stability and design of civil engineering structures, among which were railways and harbours; design of machinery; and prime movers. After passing the two stages, a Bachelor of Science degree was awarded. Once an engineer had completed his pupilage and was over twenty-one, he could become a Doctor of Science by taking two further examinations. Candidates for this higher degree were obliged to choose a practical engineering paper, either the design of machinery, or designing and estimating in civil engineering; and also an applied science paper, from applied mathematics, chemistry, geology, a branch of natural philosophy, or telegraphy.[112] For undergraduates, Jenkin offered additional courses where necessary, for example in mechanical drawing: 'I have actually tried to introduce that myself to the university, though

---

[108] *Samuelson committee*, p. 131.
[109] *Samuelson committee*, p. 131.
[110] *Samuelson committee*, p. 131.
[111] *Devonshire commission*, p. 375.
[112] *Devonshire commission*, p. 332.

I regard it as belonging more properly to elementary or second grade education'.[113] He had found that, like Tait, he had to include some basic mathematics: 'Just as Professor Tait has to teach trigonometry, so I am obliged to teach for a portion of my course statics and dynamics'.[114] Tait had introduced a class in mathematical physics in 1868 to support the new engineering degree, a course also open to his natural philosophy students.[115] As well as the engineering class, Jenkin taught a winter course in mechanical drawing on three days a week, and also gained permission to offer surveying and levelling from the summer of 1869.[116]

Jenkin's educational model could not be imposed in its entirety upon the University of Edinburgh. To change practice in the wider world – particularly the system of pupilage – was beyond his influence. There were other constraints. Despite Baxter's endowment, teaching facilities remained even worse for engineering than for science. Jenkin's model aimed to produce a high standard of professional engineer, one capable of filling senior positions, yet this demanded an investment of time and money in education which many of his students could ill afford, and which was perhaps beyond the ambition of most. The majority of students never entered the examinations for the Bachelor of Science. The degree in civil engineering as a branch of science was approved by the Senate in March 1870, but the first awards were not conferred until 1873.[117] There were no doctorates in engineering until Archibald Campbell Elliott in 1888, after Jenkin's death. Alfred Ewing became only the ninth person to graduate, in 1878 at the close of the engineering department's first decade, and by 1888 there had been in total only thirty-eight graduates.[118]

Jenkin's insistence upon a matriculation examination was soon modified, so that students could join the course without specific qualifications, although they risked being sent down in the first week if he discovered they were ill-prepared. Later it seems that Jenkin relented even on this, for he came to understand that poor grounding was not the fault of students, as few schools could equip them for his course.[119] 'The second grade schools are marvellously bad as regards the teaching of science'.[120] The course had to be fitted into two university sessions, and had to prepare undergraduates for all branches of engineering. Jenkin's contract stipulated that he should 'teach and instruct students in the science of engineering both civil and mechanical' and apply that

---

[113] *Devonshire commission*, p. 101.
[114] *Devonshire commission*, p. 101.
[115] Edinburgh University Library, *University Court Minute Book*, I, 27 April 1868.
[116] Edinburgh University Library, *Senate minutes*, 9 April 1869 and 27 November 1869.
[117] Edinburgh University Library, *List of graduates 1859–88*, p. 135.
[118] Edinburgh University Library, *List of graduates 1859–88*, pp. 133, 135. See also Emmerson 1973, p. 124 and Birse 1983, p. 104.
[119] See Sanderson 1972, ch. 5, especially p. 151.
[120] *Devonshire commission*, p. 102.

science to manufacturing industry.[121] As students who were to become professional engineers would serve a pupilage afterwards, there was no point in spending the university years on practical training, and Jenkin abided by the philosophy he had explained to the Samuelson Committee, of a scientific foundation with some transferable skills.

Even so it seems surprising that Jenkin did not include his own speciality, telegraphy, in his teaching. He had had plenty to say about the inadequacies of attempts by the government to train and examine telegraph engineers. The Royal Engineers Institute at Chatham, where he sometimes lectured, recruited 'young men of remarkable ability, selected by examination, but they did not know a tithe of what was known by my assistants ... if I had required to send out a man to a given station abroad to look after things, I should have taken one of my own assistants in preference to those men'.[122] He had also poured scorn on the test for engineer appointments to India:

> I think that examination is precisely an example of what an examination ought not to be; the examination upon theoretical subjects, such as algebra and geometry, is of the very feeblest kind, requiring merely a little knowledge of plane geometry, and very little algebra indeed; and then the examination upon practical subjects is as bad upon quite other grounds. I do not know what kind of answers they get at these examinations, but I know that some of the questions which are asked of students, and which they required to answer at once, without having any means of reference at hand, are precisely of such a nature that if they were given abroad the student would receive a month, or perhaps two months, to answer them in.[123]

There was a more general problem in finding adequately trained telegraph engineers for foreign postings:

> The men who are now being sent out upon very large salaries to different stations that I know of are, I consider, incompetent; they are incompetent to test cables, to find out whether cables are in a good or bad condition. Their theoretical requirements are far below what I think necessary. I cannot find men to recommend for those appointments.[124]

While Jenkin had never been averse to introducing highly topical subjects to his students, for example in delivering a paper on telpherage in 1883,[125] it

---

[121] Edinburgh University Library, *Senatus minutes*, 26 June 1868.

[122] *Devonshire commission*, p. 94.

[123] *Samuelson committee*, p. 132.

[124] *Devonshire commission*, p. 101.

[125] Hempstead 1991, p. 144.

was not until after his death that a course of twenty lectures in electrical engineering was introduced in Edinburgh, by his successor George Armstrong.[126] Perhaps significantly, Armstrong's course was not repeated, maybe recognising the difficulty of staying abreast of developments in the field. Telegraph engineering may also have been seen as only marginally relevant in Scotland, as the cable-making and -laying industry was concentrated upon the banks of the Thames in London. Many of Jenkin's Scottish students expected to enter more traditional branches of civil or mechanical engineering, so that it was sensible to give priority to material which would be useful to them. Jenkin's star student, Alfred Ewing, destined to become an eminent telegraph engineer, thus began his university career by writing Jenkin an essay on the relative merits of wet and dry sewerage systems.[127]

It was an auspicious time to enter academic engineering. With a growing public and political interest in the education of engineers, Jenkin did not face the open hostility which Lewis Gordon had encountered at Glasgow University in the 1840s and early 1850s, from colleagues and from the university authorities which had denied him a lecture room and demonstrating facilities.[128] Nor did he have Gordon's problem attracting students.[129] Twenty-nine students joined Jenkin in 1868, and from 1869 when the course was fully running, total numbers stabilised at about fifty.[130] Most attended the winter sessions, with fewer students in the summer surveying class. Robert Louis Stevenson attended one of these in June 1870, a levelling course in the Braid Hills, when the reluctant engineering student managed to run a levelling rod into his leg.[131]

Stevenson, usually ill-disposed towards his engineering studies, nonetheless enjoyed Jenkin's class excursions for their 'tavern merriment in the free air of Heaven'.[132] Jenkin had taken up Baxter's idea for practical excursions which would introduce students to different kinds of engineering establishment. This seems to have been unusual among engineering departments, for in the words of Stevenson, 'this form of iniquity is perhaps peculiar to Edinburgh'. Pushed unwillingly towards the family tradition of service to the Northern Lighthouse Board, Stevenson, described by Alfred Ewing as 'a nominal student of engineering',[133] left an account of one of Jenkin's industrial visits which took place at the end of the university session in April 1871. The engineering class

---

126 Birse 1983, p. 101.
127 Bates 1946, p. 3; Ewing 1933, p. 259.
128 Birse 1983, p. 64.
129 Birse 1983, pp. 64–65.
130 *Devonshire commission*, p. 335. Numbers from 1869 to 1874 ranged between 45 and 55. The university itself had four faculties and 1400 students in 1868: Edinburgh University Library, *University Court Minute Book*, I, 3 April 1868.
131 Booth and Mehew (eds) 1994, I, p. 200.
132 National Library of Scotland, Acc.9690.
133 Ewing1933, p. 249.

had organised a drunken end-of-year supper, with revelries continuing until four in the morning, after 'many consecutive hours of rum punch'. The class then had to assemble at seven at Waverley Station, for a third-class rail journey to a mechanical engineering factory outside Glasgow. 'We were a somewhat wretched looking party. One fellow had not been home at all, and appeared in the crumpled dress clothes of the evening before, like a moon at midday'. Still, continued Stevenson:

> class excursions are crucial tests of a student's abilities. Unless he can attain the ideal combination of the cheap excursionist and the rowdy in his walk and conversation, he is not thought to be in form at all.[134]

Undeterred by the 'languor and weariness' of his group of crapulent students, Jenkin pressed on towards the Glasgow factory. 'I felt an unreasonable anger towards our Professor who, bag and water-proof on shoulder, danced down the road with a vitality that was little short of insulting.' concluded Stevenson.[135]

While Stevenson's account aimed mainly for dramatic effect, it can be divined from this story that Jenkin applied his customary enthusiasm to the industrial visits. There is no suggestion that new educational ideas had to be forced upon him. His rival for the Edinburgh chair, Rankine, successor to Gordon in Glasgow in 1855, was respected for his systematic engineering course, yet had lost out to Jenkin, whose new employers were evidently attracted by his open mind and practical approach.[136] Jenkin has been compared unfavourably with William Ayrton as a progressive educator,[137] but the two were not strictly contemporaries and Jenkin enjoyed less freedom to try out new educational methods. Ayrton, half a generation younger, launched a telegraphy course in Japan with the advantages of a clean slate and good laboratory facilities, and unencumbered by an inhibiting framework of pupilage. He was also dealing purely with electricians, unlike Jenkin who was required to educate the whole range of professional engineers.[138]

Jenkin was widely admired as 'a highly skilled and enthusiastic teacher, who was thorough and systematic in everything that he did ...'[139] To his students, 'the work of the engineering class was absorbing and exacting'.

---

[134] National Library of Scotland, Acc.9690.

[135] National Library of Scotland, Acc.9690.

[136] Birse 1983, pp. 65–66.

[137] Gooday 1991. For details of the career of Ayrton (1847–1908) see the *Dictionary of National Biography*.

[138] See Gooday 1991, pp. 85–90.

[139] Obituary, *Journal of the Society of Telegraph Engineers*, p. 347.

His teaching was highly successful, original in conception, and altogether characteristic of the man. Seizing the salient points of his extensive subject, he passed from one to another with a rapidity that left little breath in the boy fresh from the country, and used to the drudgery of school drill. But if fast, the teaching was clear and thorough, shirking no difficulties that were natural to the subject, and presenting none that were artificial; and the boy, if he was good for anything, was bound to follow, though panting. Each part in turn was illustrated by a host of numerical examples – practical questions set for solution at home in forms drawn from the teacher's experience. These at once clinched the teaching of the lecture–room, and accustomed the lads to look on the theory they were learning as no mere abstruse piece of mental gymnastic, but as a tool for daily use in the drawing-office and the workshop in after life.[140]

This is clear confirmation of Jenkin's ability to bridge the worlds of theory and practice, between academia and the workshop. He was also able to make engineering accessible to students whose understanding of mathematics was limited:

A notable feature of his style was the comparative absence of mathematical form. He had himself a distaste for algebra, which led him to employ it sparingly, except in the simplest applications; but he managed to do this with little if any sacrifice of logical completeness or precision in his exposition of engineering theory. In later days some of his pupils came to realise that they had all the while been learning mathematics without knowing it.[141]

This has been misread by one author as implying that Jenkin had an ingrained distaste for mathematics.[142] In fact, although modest about his own mathematical aptitude, Jenkin was not without ability and continued to apply himself to the subject, even after he had left Fairbairn and Dr Bell. He had written to Thomson in 1859 from Birkenhead: 'I am working at the higher calculus and do not despair of managing it'.[143] Jenkin was especially attracted to graphical methods, which he was responsible for introducing to the study of political economy as well as using them in the more conventional context of science and engineering.[144] He also maintained a general interest in the teaching of mathematics – writing to Stevenson in 1883, apparently in response to a request from the author to recommend a tutor in mathematics and physics, he claimed that 'coaching in these subjects is much better done in France, though

---

[140] Obituary, *Institution of Civil Engineers Proceedings*, p. 368.
[141] Obituary, *Institution of Civil Engineers Proceedings*, p. 368.
[142] Birse 1983, p. 98.
[143] Kelvin, Glasgow, J13, Jenkin to Thomson , 4 August 1859.
[144] For Jenkin's contribution to economics, see Chapter 6, below.

the French are awfully behindhand in the higher walks – also much better in Germany'. He added that 'a very clever young man can teach himself in Gt. Britain perhaps better than anywhere else'.[145]

Although the outline syllabus varied little while Jenkin held the Edinburgh chair – he offered the same mix of engineering and drawing in winter, surveying in summer – the content was constantly updated. Lectures were frequently rewritten in line with developments in the field, or with his progressing thoughts:

> In Jenkin's hands the work of the class was certain not to fall into set grooves. His teaching was no easy repetition from year to year of stereotyped lectures. A book or paper would suggest a new mode of presentation, or the idea would occur without tangible inspiration, and immediately he would write out afresh a whole section of the course with all the zest of a novice. When ... *Matter and Motion* by Clerk Maxwell was published, it took possession of Jenkin, dominating him for days, and under its inspiration, he remodelled his whole treatment of dynamics.[146]

Alfred Ewing considered Jenkin to have been 'the most inspiring teacher he had ever known' and furthermore 'one of the kindliest and friendliest of men'.[147] Both Jenkin and his wife took an interest in talented students, and they grew particularly close to Ewing, who was later a trustee of Jenkin's estate and co-editor of his collected papers.[148] In 1920 Anne Jenkin wrote to Ewing of the fifteen years that she and Jenkin had 'delighted over you together', and the thirty five years of enduring friendship after Jenkin's death.[149] This generosity towards his juniors was noted by others:

> He was an enthusiastic admirer of ability in other men, and he was especially warm in his encouragement of beginners, whether they were his own pupils or not. To gain his help it was only necessary to let him see that it was anxiously wished for, and that the recipient was not likely to make a mean use of it.[150]

> The hand held out to younger men on lower rungs of the ladder – the anxiety to secure for any who worked with him the fullest share of credit, the delicate little conspiracies in which he plotted with his wife to do

---

[145] Yale, Beinecke 4990, Jenkin to Stevenson, 3 August 1883.
[146] Obituary, *Institution of Civil Engineers Proceedings*, p. 368.
[147] Ewing 1939, pp. 84–85.
[148] Ewing 1939, p. 84.
[149] Ewing 1933, p. 255.
[150] Obituary, *Nature*, 18 June 1885, p. 155.

some unlooked-for kindness to those they loved, less prosperous than themselves – these are things which it is good to have seen and known.[151]

While work experience could not be a formal part of the degree programme, in the way that Henry Dyer in Japan was able to introduce a complementary summer pupilage between academic sessions,[152] Jenkin could employ a few students on summer consultancy work. This gave them a practical education of the kind which he believed could never be replicated in a laboratory. It also gave Jenkin and his associates an opportunity to try out potential assistants. There was a chronic shortage of such young men in Britain, and in 1883 Jenkin described his consultancy partnership's continuing difficulty in finding them:

> Within the last fortnight I was required along with Sir William Thomson to find a young electrician to send out upon a mission, and we were extremely puzzled where to turn. I found one man, and if I had not found him I don't know where I could have gone for another ... In the work we are carrying on I have had to employ a foreigner, and I am often unable to find the person that I want.[153]

Thomson as well as Jenkin had used a number of 'protégés' on the French expedition.[154] William Ayrton himself fell into this category, studying under Thomson and later working as a junior employee for Thomson and Jenkin.[155] After Jenkin's death, Forde turned to Thomson for help in finding new young employees. 'Could you recommend us a good electrical assistant in case we should require one by and by ... we also have room for a pupil should you hear of one.'[156] Training and recruitment was arranged through this interchange between academic and working engineers, rather in the way that Jenkin himself had been recommended to Newall by Gordon. Jenkin had long acknowledged, not altogether approvingly, the importance of patronage – 'connections with men who find the money' – in gaining employment.[157] With an extensive knowledge of the engineering network and his ability to recognise talent, as professor Jenkin used the system to advantage in recommending and placing his best students. 'He gradually brought around him a number of able

---

[151] Obituary, *Institution of Civil Engineers Proceedings*, pp. 376–77.
[152] Gooday 1991, p. 86.
[153] *Educational Endowments (Scotland) Commission*, p. 787.
[154] Kelvin, Cambridge, A35, James Anderson to Thomson, 1 June 1868.
[155] Gooday 1991, pp. 76–77.
[156] Kelvin, Cambridge, F254, Forde to Thomson , 21 January 1886.
[157] *Samuelson committee*, p. 137. This point was also emphasised by Jenkin in his advice to the young Forbes: University of St Andrews Library, *Forbes correspondence* (in) 1865/62, 5 July 1865.

and accomplished young men, much of whose professional success in life is directly traceable to his influence'.[158]

Alfred Ewing had no doubt of his debt to Jenkin. He had arrived at the university in 1871, as 'a youth of sixteen, who brought no introduction'.[159] At the end of that first session, Ewing was prizeman in the class of Engineering, and Jenkin surprised him with the offer of a place on the staff of his and Thomson's consultancy company.

> For a young man without influence or prospects this was an opportunity not to be missed. I went at once to London at their bidding, and set about learning how to make electrical tests in the cable factory, which at that date was a better school of electricity than any laboratory ...[160]

He then returned for a second university session, and won the class medal, followed by another summer at Hooper's Telegraph Works on the Isle of Dogs and three separate cable-laying expeditions to Brazil. In 1874, at nineteen, Ewing was considered sufficiently experienced to represent Jenkin and Thomson on a south American expedition.[161] Not until 1876 did he return to university, as Jenkin's assistant on various pieces of scientific and engineering research. Much of this work took place at Jenkin's house, and during this period Ewing became an intimate of the family. In 1878, six years after going up to Edinburgh, and on the eve of his departure for Tokyo as Jenkin's nominee for the chair, Ewing finally took his degree.[162]

While Alfred Ewing was the greatest star in Jenkin's firmament, there were other notable *alumni*. The engineering department at Edinburgh produced several Whitworth Scholars in Mechanical Engineering, and also C. Michie Smith, who became professor at the Sir Josiah Mason College, Birmingham.[163] One of Jenkin's first graduates, Charles Stevenson (1855–1950), a cousin of Robert Louis, took over the family business in lighthouse engineering and made a mark as a pioneer of wireless telephony.[164] Jenkin's middle son, Charles Frewen Jenkin (1865–1940), who took prizes in 1884 and 1885, went on to hold the first chair of Engineering Science in the University of Oxford in 1908. The youngest son, Bernard Maxwell Jenkin, won a second class prize in 1885–6, and in 1889 became partner in a firm of consulting engineers with Alexander Kennedy, who resigned his engineering chair at University College, London to

---

[158] Obituary, *Journal of the Society of Telegraph Engineers*, p. 347.

[159] Ewing 1933, p. 257.

[160] Ewing 1933, p. 257.

[161] Bates 1946, pp. 2–3; Ewing 1939, pp. 25–29.

[162] Birse 1983, p. 104; Ewing 1933, p. 258.

[163] Obituary, I *Journal of the Society of Telegraph Engineers*, p. 347

[164] On the Stevenson family as engineers, see Mair 1978.

become Jenkin's partner.[165] The influence of Jenkin's department spread beyond Scotland as undergraduates came from around the world, from India and Australia, Japan and the United States.[166] And Jenkin took other young men under his wing, from beyond the universities of Edinburgh and Glasgow. One was Andrew Jamieson (b.1849), a graduate of Aberdeen who worked for Thomson and Jenkin from 1873 on the south American projects;[167] another Sidney F. Walker (b.1852), with a background as a mathematician, naval officer and student of telegraphy, who was considered fortunate to gain employment with Thomson and Jenkin at Hooper's Telegraph Works, and was later a distinguished mining engineer.[168]

Some of the more impoverished Victorian professors wrote textbooks to augment their income. Sir William Thomson viewed this with disdain, likening the practice to 'an army in which the general is employed in teaching the goose step to recruits'.[169] Jenkin was no longer short of money, but nor was he as proud as Thomson. In writing the book *Electricity and Magnetism*, published by Longman in 1873, he was evidently driven by a mission to systematise and explain.[170] Jenkin, it is said, recognised that telegraphy was built upon fragments from the work of various scientists, with the result that the subject was framed in a jargon comprehensible to telegraphers yet foreign to academics. The difference had arisen because of the need for quantitative results in telegraph-circuit operations.[171] His was the first book on electricity written with students in mind, in a modern style, and was a huge success. It passed through many editions, was translated by Jenkin into German, French and Italian, and despite rapid advances in the subject remained the standard textbook at the time of his death. It was said to have 'marked a new departure in the exposition of the subject as a quantitative study'.[172]

It was, in fact, the first systematic exposition of that new electricity to which the labours of the British Association committee had given form, the science of the practical electrician, which was, as Jenkin himself said, more scientific than the science of the older text-books.[173]

---

[165] Kennedy and Jenkin later changed its name and became the well-known engineering firm Kennedy and Donkin.

[166] Edinburgh University Calendar, engineering prize lists, for example 1874–75.

[167] *Electricians' Directory*, 1885, p. 108.

[168] *Electricians' Directory*, 1885, p. 133.

[169] Morrell 1976, p. 84.

[170] His Electricity, published in the SPCK Elementary Manuals of Science series in 1881, was in the same vein.

[171] Appleyard 1939, p. 69.

[172] Jenkin's entry in the *Dictionary of National Biography*, XXIX (1892), p. 296.

[173] Obituary, *Institution of Civil Engineers Proceedings*, p. 374. See below for Jenkin's work on the British Association for the Advancement of Science (BAAS) committee.

Another obituarist, looking back at the impact of Jenkin's innovative approach, confirms Hunt's view that Thomson , Jenkin, Maxwell and others, besides extending the physics laboratory into the cable industry, were also responsible for bringing part of the cable industry to the laboratory[174]:

> Jenkin's book ... was a revelation to non-mathematical and even to many mathematical men, of the ideas which had until then been wrapped up in the mystery of mathematics or in the practice of the submarine cable testing-rooms. Sir William Thomson had been publishing many detached papers on electricity in the mathematical journals, and had been applying his knowledge in practice, so that an exact science of electrical quantities had been growing up among submarine cable engineers; but the electricity of the text books remained as unscientific and primitive as of old: the knowledge of the practical men had become indeed far more scientific than the knowledge of the schools.[175]

To present electricity and magnetism as subjects capable of quantitative study was an entirely new departure. It must be remembered, says Jenkin's obituary in *Nature*, that 'at that time 'electric potential', which today has its commercial unit, was to everyone, except the engineers of submarine telegraphy, a mere mathematical function.' Jenkin was modestly pleased with his popular books, writing to Thomson in 1881 that a fifth edition of *Electricity and Magnetism* was out, 'and a little shilling book besides' – presumably referring to the Society for the Propagation of Christian Knowledge edition of *Electricity*.[176]

The writing of textbooks is consistent with Jenkin's development as an educator and communicator. He was also engaged upon more weighty publications, for the university post enabled him to develop the scientific and technical investigations which had been his central interest from the time of his work with Newall. Jenkin's growing scientific reputation, even before he had held an academic post, was reflected in his election as a Fellow of the Royal Society in 1865. In 1863 Thomson wrote to Sir George Gabriel Stokes for advice on Jenkin's application for FRS: 'He certainly is very zealous for science, & has already done good work besides having proved remarkable ability'.[177] The following year, Thomson wrote again to Stokes: 'He is remarkably clever and eager in the pursuit of science, both in acquiring knowledge himself, and in endeavouring to obtain new results by experiment ... Altogether Jenkin is a most useful man for science, and a remarkably sound

[174] Hunt 1994, p. 63.
[175] Obituary in *Nature*, 18 June 1885, p. 153.
[176] Kelvin, Cambridge, J29, Jenkin to Thomson, 15 September 1881. *Electricity and Magnetism* reached a tenth edition in 1891.
[177] Wilson (ed.) 1990, I, p. 315, Thomson to Stokes, 3 December 1863.

worker both in pure science and in engineering applications'.[178] Jenkin's other supporters included Rankine, Fairbairn, Douglas Galton and Maxwell, and his most significant work to date was summarised thus:

[Jenkin] made the first direct measurement of the specific resistance of any so-called insulator, as compared with that of conductors ...(1859); the first published measurement of electric absorption by dielectrics, producing an apparent extra resistance to the voltaic current ... (1859–60); the first published measurement of the specific inductive capacity of dielectrics, by means of voltaic currents ... (1862)[179]

Thomson considered that the first of these, as an accurate determination of a property of matter in approximate absolute measure, was the most valuable. Jenkin's then most recent work, said Thomson, 'contains remarkable verifications of the mathematical theory, besides supplying the numbers required for reducing the results to absolute measure for speed of signalling'. Thomson believed it to be 'very interesting and valuable with reference to practical signalling'.[180]

At first Jenkin had been diffident about publishing his research, but soon realised that he could play a part in restoring confidence in deep sea cables. Even in 1860, he had little doubt that they would work, and he wrote to Galton:

I had no intention of publishing anything on [faults in submarine cables] before making a complete course of experiments. My conversation with you, however, showed me that even the few facts of which I am in possession may be interesting, and may perhaps in some measure restore confidence in the durability of submarine cables, by showing that recent failures probably depend on causes which, if known, can be avoided.[181]

The confidence of Thomson and Jenkin that problems would be solved through a development of theory, the belief that success could be based only upon science, was reassuring for company directors who wanted to attract investors. Yet inevitably these same commercial interests pressed the scientists for 'quick fix' solutions which were difficult to reconcile with the rigorous standards which Jenkin imposed upon himself. The academic appointment enabled a more detached view, for while still fully in touch with commercial electrical engineering through his consultancy work, Jenkin could remain aloof from some of the tension between that world and the scientific objectives of his university life.

[178] Wilson (ed.) 1990, I, pp. 319–20, Thomson to Stokes, 20 April 1864.
[179] Royal Society, Printed List of Candidates, 1848–67.
[180] Wilson (ed.) 1990, I, pp. 319–20, Thomson to Stokes, 20 April 1864.
[181] Galton committee, p. 145. Jenkin wrote from Turin, 28 September 1860.

Over the course of his career, Jenkin published about forty scientific papers, mainly on electricity and its applications.[182] Thomson was the first to suggest to Jenkin that some of his work should be published, and he encouraged and advised the younger man in his earliest efforts. Jenkin had to seek his employer's approval – 'I hope [Mr Gordon] will have no objection to my writing a paper as you propose. I think I have materials which would make it interesting & I do not think any of Messrs R.S. N[ewall] & Co's rivals would be much benefited by the information divulged'.[183] The company agreed that Jenkin could communicate his experiments through Thomson to the Royal Society or to the British Association for the Advancement of Science. Jenkin planned to write on the speed of signalling in submarine cables, and on sending codes. He asked Thomson to advise him: should both papers be sent to the Royal Society or one to the British Association? Was a full or concise version required? When should he send it? And how could he obtain Thomson's papers on electricity?[184] It proved to be more of a challenge than he had expected: 'I am very busy trying to make a good paper but there is more work than I had anticipated'.[185] He sent one of the two papers to Thomson in September 1859, and handed over some editorial control: 'You can of course omit any part of it which you do not think 'the thing''.[186] Whether these were credited to Thomson is unclear, but Jenkin anticipated making a wider reputation. He asked Thomson: 'In giving my name may I ask that you will give the 'couple' [Fleeming Jenkin] as they form a good mark of identity – my surname being in no wise remarkable'.[187]

In February 1860 Jenkin told Thomson that he had been busy 'finishing the paper on gutta percha'.[188] There may then have been a delay in his publishing plans, for he was warned by Galton not to submit any papers to 'philosophical journals' before the Joint Committee had published its findings.[189] In 1861 Jenkin in his turn helped Thomson, who was laid up in Glasgow with a broken leg, by ensuring that Thomson's papers on the age of the sun's heat and the internal heat of the Earth found their way on to the agenda of the British Association for the Advancement of Science meeting in Manchester.[190] Jenkin himself gave a paper at that conference, and reported to

[182] Obituary, *Institution of Civil Engineers Proceedings*, p. 374.
[183] Kelvin, Glasgow, J12, Jenkin to Thomson, 30 July 1859.
[184] Kelvin, Glasgow, J13, Jenkin to Thomson, 4 August 1859.
[185] Kelvin, Glasgow, J14, Jenkin to Thomson, 30 August 1859.
[186] Kelvin, Glasgow, J17, Jenkin to Thomson, 9 September 1859.
[187] Kelvin, Glasgow, J18, Jenkin to Thomson, 13 September 1859.
[188] Kelvin, Glasgow, J32, Jenkin to Thomson, 3 February 1860. This was 'On the insulating properties of gutta percha', published in *Proceedings of the Royal Society*, X (1860).
[189] Kelvin, Glasgow, J40, Jenkin to Thomson, 27 August 1860.
[190] Morris 1994, pp. 325–26. Jenkin was also interested in heat, having suggested a new method of heat measurement to Thomson in 1860, eleven years before Siemens' publication of

Thomson that it had gone off very well, 'attracting a good deal of attention. Nothing was said at its close but in the Committee next day Airy [the Astronomer Royal] and others spoke very satisfactorily and voted on Joule's resolution that the Committee of Research be requested to give me seventy pounds for further research'.[191] He had also tried to show an experiment, but without much success as the room could not be darkened, so that he later gave a small private demonstration to Joule, Wheatstone and others.[192]

In the years from 1859 until his move to Edinburgh, Jenkin had at least a dozen scientific papers or lectures published, in journals including the *Proceedings* of the British Association for the Advancement of Science, the *Philosophical Magazine* and the *Proceedings of the Royal Society*. Later he published in the *Transactions of the Royal Society of Edinburgh, Journal of the Society of Arts, Nature*, and the *Journal of the Society of Telegraph Engineers*. He also willingly wrote for less prestigious journals which reached a far wider audience, such as *Good Words*, in which he had an article on telpherage shortly before his death.[193] Besides Thomson, Jenkin consulted Maxwell and Tait for advice on articles in preparation. To Maxwell in 1868 he sent 'part of a revise of my atoms article' with a request to 'write anything you please in pencil on the margin!'.[194] Thomson and Tait had also seen the article, for Jenkin replied to Thomson: 'I have not made the elementary blunder about impact which you and Tait attribute to me, and I pretend that if you read my assertions carefully with the context they are perfectly correct (Maxwell passed it)'.[195] Jenkin did concede that comments about relative motion needed clarifying to prevent misunderstanding. When thanking Maxwell for annotations on part of his article on Lucretius, Jenkin implied that he found Tait's criticisms harder to take:

Tait has been down upon me for a series of blunders and oddly enough they are none of them the same as your hits[?]. If I had time (and spirits)

---

the same method of thermometry: Wilson (ed.) 1990, II, pp. 362–63, Thomson to Stokes, 7 October 1871.

[191] Kelvin, Cambridge, J37, Jenkin to Thomson, undated but presumably September 1861. The reference is to James Prescott Joule (1818–89).

[192] Kelvin, Cambridge, J37, Jenkin to Thomson, presumably September 1861. Sir Charles Wheatstone F.R.S., (1802–75), had been a pioneer of overland telegraphy during the 1830s.

[193] Obituary in *Nature*, XXXII, 18 June 1885, p. 154. Jenkin's publications are listed in the bibliography.

[194] Cambridge University Library, Add 7655/II/28, Jenkin to Maxwell, 10 [January?] 1868.

[195] Kelvin, Cambridge, J27, Jenkin to Thomson, 20 February 1868.

would calculate how many annotators it would take to show me that the whole is wrong from beginning to end.[196]

'Lucretius and the Atomic Theory', a review of the second edition of H.A.J. Munro's *Lucretius*, was published in the *North British Review* in 1868.

Jenkin's interests in Lucretius and the ultimate constituents of matter reflect a concern with metaphysics and epistemology which developed from his early involvement in electrical theory and measurement.[197] Most assessments of Jenkin's life agree that his most enduring achievement resulted from work on electrical standards which started early in his career, before he was thirty. In 1861, at the Manchester meeting of the British Association where Jenkin had helped steer Thomson's papers on to the agenda, the absent Thomson had also written to suggest a Committee on Electrical Standards, and propose Jenkin as reporter or secretary. The idea was approved, Jenkin told Thomson, Wheatstone having

> framed a resolution, not altogether a bad one, and the Committee is appointed and has ... £50 for its use. You and I are to redetermine Weber's unit; compare it with the one he will furnish – Matthiessen and Williamson are to try the permanency of wire and its reproductibility (my word) – Today Sir C. Bright and Latimer Clark have come down and want to get on the Committee ... The General Committee is I think yourself, Wheatstone, Miller, Matthiessen, Williamson and myself. I hope Sir C. Bright and Latimer Clark will get on – they seem to have found out the necessity of a dependence between the various electrical units and had even prepared a paper on the subject but I got hold of them and pointed out that it was all done already and I must say Latimer Clark looked delighted and is eager to have it all explained.[198]

Bright and Clark tabled their paper, 'On the Formation of Standards of Electrical Quantity and Resistance', which was to be referred to Jenkin's committee.[199] Latimer Clark was later to claim that their proposal for a 'system of perfectly inter-dependant units founded on metrical measures and on an Electro-static base' was the earliest such suggestion and was 'almost identical with that which has been finally adopted and is now in use, the Volts, Ohms, Farads &c. having the same unitary correlation'. Though Bright was credited

---

[196] Cambridge University Library, Add 7655/II/51, Jenkin to Maxwell, 28 October 1871.

[197] Jenkin's ideas on this are contained in two essays 'Lucretius and the atomic theory' (1868) and 'A fragment on truth' (1884), which are explored further in the appendix to this chapter.

[198] Kelvin, Cambridge, J37, Jenkin to Thomson, presumably September 1861.

[199] Hunt 1994, p. 58. The paper was published in the *BAAS Report, XXXI*, (1861), pp. 37–38.

with joint authorship, Clark added, 'the original ideas emanated from me', although conceding that he 'was not mathematician enough to see the enormous value of an absolute system, founded on mass, time & space', which had by then gained the British system of electrical measurement world-wide acceptance.[200]

The standards committee met between 1862 and 1869, producing six major reports which were largely the work of Jenkin.[201] The group was charged with determining a convenient unit of resistance, and 'the best form and material for the standard representing that unit'.[202] An obituary in the *Journal of the Society of Telegraph Engineers* is unequivocal in describing Jenkin's central role in this work:

> [Jenkin] was closely associated with Clerk Maxwell, and other able experimental investigators, in carrying out the work of the committee. For many years he recorded the work of the Electrical Standards Committee (much of it his own) ...[203]

In Thomson's own words of 1864, supporting Jenkin's application to the Royal Society, there is confirmation that Jenkin and Maxwell were carrying along the main experimental work of the committee:

> Jenkin has worked hard for the units' committee in this most important and difficult matter, & along with Maxwell, and B[alfour] Stewart, have [*sic*] already gone through the work of measuring resistance in absolute electrostatic measure, with a degree of accuracy much superior to that attained by Weber.[204]

After Jenkin's death, Thomson again acknowledged that Jenkin and Maxwell had carried out the most important of the committee's experiments.[205]

The standards committee, as Hunt has shown, was initiated by Thomson, and not as sometimes claimed by Bright and Clark, who became members of it only in 1862 and 1866 respectively.[206] Thomson was apparently responsible for drafting much of the first report, in 1863, and the committee's considered view that the ohm should be an absolute unit of resistance was in accordance with his

[200] Kelvin, Cambridge, C91, Clark to Thomson, 3 May 1883.
[201] Chipman 1973, p. 93.
[202] Smith and Wise 1989, p. 687.
[203] Obituary, *Journal of the Society of Telegraph Engineers*, p. 346.
[204] Wilson (ed.) 1990, I, pp. 319–20, Thomson to Stokes, 20 April 1864.
[205] W. Thomson, 'Obituary Notices of Fellows Deceased', *Proceedings of the Royal Society*, XXXIX (1885), p. ii.
[206] Hunt 1994, p. 58.

own views.[207] Indeed, Thomson later wrote that the committee had been set up expressly 'for the purpose of promoting the practical use of Gauss and Weber's system of absolute measurement'.[208] The committee established the system of ohms, amps and volts which is still in use, prepared materials for the construction of reliable resistance units, and developed methods of measuring resistance to a precision of 0.1 per cent. In 1867, Jenkin made the first absolute measurement of capacitance.[209]

Latimer Clark had expressed worries about the committee's composition, claiming that they were people 'but little connected with practical telegraphy'.[210] Yet the committee, as Hunt points out, wanted to serve the needs of the cable industry and knew that their system would not be acceptable unless it satisfied telegraphers. The official resistance standards were announced by Jenkin in the *Philosophical Magazine*, in February 1865, in the form of a unit coil and box priced at two pounds ten shillings. The new standard ohm quickly found its way around the globe, to major telegraph companies, foreign governments and colonial telegraph departments, as well as physics laboratories and instrument makers.[211]

This project assumed a significance far beyond its apparently dry academic remit, for it produced a means of reconciling practical electricians with theoreticians, and in doing so elevated electrical engineering to a scientific discipline.

> It would be hard to overestimate the value of this work. By establishing practical units in which electrical magnitudes could be expressed, the committee did much to bridge the gulf which had, up to that time, separated the electricity of the schools from the electricity of the cable factory and the telegraph station. It gave a common standing ground to the mathematician and the engineer, who at once found themselves of mutual advantage. It stimulated the purely scientific side of the subject by enriching and simplifying its vocabulary, and by placing in the hands of physicists new instruments and methods of research. It raised electrical engineering from empiricism to the rank of applied science; and it paved the way for that rapid progress in the applications of electricity which is so marked a feature of the present time ...[212]

---

[207] Smith and Wise 1989, p. 688. There is some controversy over the authorship of the reports. Smith and Wise believe that Thomson was responsible for some parts – see p. 688, footnote 10 – while Hunt emphasises Jenkin's role: 1994, p. 60.

[208] W. Thomson, 'Obituary Notices of Fellows Deceased', *Proceedings of the Royal Society*, XXXIX (1885), p. ii.

[209] Chipman 1973, p. 93.

[210] *The Electrician*, 17 January 1862, quoted by Hunt 1994, p. 59.

[211] Hunt 1994, pp. 60–61; Hunt 1997, p. 325.

[212] Obituary, *Institution of Civil Engineers Proceedings*, p. 366.

Between 1881 and 1884, the recommendations of Jenkin's committee were broadly adopted by the International Congress on electrical standards in Paris.[213]

Jenkin's disposition suited him for the electrical standards project. In the first place, he was a man with an overview. His approach was holistic yet fundamental, his interest excited by a task which created channels of communication between theory and practice in electrical engineering, in the way that his text book later also attempted to reconcile the two traditions. He did not evade the confrontation with philosophical and epistemological problems which the electrical standards role demanded. His evidence to Galton shows Jenkin already holding a wide concept of submarine telegraphy, recognising the need to standardise units of measurement and terminology in electrical engineering, so that disputing electricians could at least argue in a common language.

The other aspect of Jenkin's character which disposed him towards the work on standards was his keen interest in measurement. His earliest correspondence with Thomson was on how best to measure the resistance of copper and gutta percha, and resulted in the paper submitted to Galton's committee, 'On the Insulating Properties of Gutta Percha'. Measuring the specific resistivity of gutta percha on the same scale as that of copper was a challenging task given the extraordinary contrast between insulator and conductor; Jenkin himself compared it to the difference between the velocity of light and the speed of a body travelling one foot in 6700 years.[214] He attributed to Thomson the inspiration in 1857 that an insulator was actually a conductor of enormous resistance which could be placed on the same scale. Jenkin took up the suggestion of making systematic measurements, and from 1859 all undersea cables were tested on a system which determined the resistance per mile of the insulator. Jenkin produced a ratio of resistivities so that the leakage of current through any given length of cable could be calculated.[215] Measurement remained a major preoccupation. His second son, the engineer Frewen Jenkin, writing to Robert Louis Stevenson soon after Jenkin's death, remarked upon this interest: 'Most of all I have inherited his desire to measure everything, a wish which very few people seem to have, why I can't think'.[216] Late in his life, in a paper delivered at an educational conference in 1884, Jenkin stated his principle that the work of a scientific laboratory should be training in the art of measurement.[217]

---

[213] Smith and Wise 1989, pp. 694–95.

[214] 'Submarine Telegraphy', *North British Review*, XLIV (December 1866), p. 468.

[215] Hunt 1994, pp. 54–55. See M.N. Wise, *The Values of Precision*, (Princeton University Press, 1995).

[216] Yale, Beinecke, 4982, C.F. Jenkin to Stevenson, 9 August 1887.

[217] Obituary, *Institution of Civil Engineers Proceedings*, p. 373.

Like Jenkin himself, the problem of electrical measurement, and its solution, emerged from submarine telegraphy. The multilingual engineer was largely responsible for developing a new international language of electrical standards. Jenkin's achievements as a 'philosophic electrician' have a consistency and a symmetry. He measured, he taught, he pursued scientific investigations, he practised as a telegraph engineer. The move to Edinburgh in 1868 appeared to consolidate and secure all those earlier activities and interests. Yet simultaneously, while still apparently engrossed in problems of electricity, Jenkin also began to make a mark in fields beyond engineering.

Appendix: Jenkin on Truth[218]

Jenkin's interest in Munro's work on Lucretius was not confined to a mere review of the Roman philosopher-poet's ideas.[219] He chose instead to consider how those ideas would allow an examination of the 'ultimate constituents of matter'. Jenkin saw the history of science in terms of the long existence of two competing theories, namely continuity or discontinuity, plenum or atoms. He believed that Aristotelian continuity held sway until replaced by Newton's corpuscularian theory. From the end of the eighteenth century, following the writings of Boscovich, ideas of a plenum re-emerged and were strengthened by field theories developed from concepts drawn from the work of Faraday.[220] Jenkin sensed that a new consideration of 'the real tenets' of Lucretius was timely, as scientists again, 'after a long pause in the inquiry', began to attempt to explain the ultimate constitution of matter.[221] While the Greeks had made 'subtle guesses' in trying to solve this most fundamental problem, only one of their theories – that picked up and expounded by Lucretius – came near to unravelling the enigma.[222]

The antitheistic nature of atoms did not concern Jenkin, who sought to examine Lucretius's ideas in the light of modern knowledge and to show their relevance. He rewrote Lucretius's conceptions of the actions of atoms so that descriptions matched perceptions. Arguing that natural phenomena are subject to definite laws, Jenkin pointed out that 'Lucretius [failed] to perceive that definite physical laws are consistent with the existence of God'. The very

---

[218] This section is based on Hempstead 1991, pp. 123–29.

[219] Lucretius (99–55 B.C.), a contemporary of Julius Caesar, was a disciple of Epicurus and notable free-thinker whose work disappeared into obscurity after the end of the Roman Republic, until revived during the Enlightenment: B. Russell, *A History of Western Philosophy* (London: George Allen and Unwin, 1946), pp. 270–74.

[220] Jenkin refers to Boscovich in a letter to Thomson in 1867: Kelvin, Glasgow, J69, 24 June 1867.

[221] Colvin and Ewing (eds) 1887, II, p. 177.

[222] Colvin and Ewing (eds) 1887, II, p. 178.

conception of a law, he thought, suggests a lawgiver.[223] Using as examples spectrum analysis and the laws of chemical combination, he explained that they could be seen as 'simply special applications of [Lucretius's] general theorem; if matter really obeys definite unchangeable laws, the ultimate materials employed to make matter must themselves be definite and unchangeable'.[224]

The persistence of Lucretius's atoms in spite of the waxing and waning of physical objects he likened to the Law of Conservation of Energy, and showed how the concepts of *vis viva* and elasticity were implicit in Lucretius's atomism. Lucretius's acceptance of absolute space and his opinion that in the 'swerving' of atoms and their interaction lay the origins of all that we see, were not, Jenkin thought, compatible with Newtonian dynamics. This did not allow absolute space, and did not permit random uncaused 'swerving'. Jenkin appears to have required a prime mover in his universe, for Lucretius's atoms 'were described as in deadly stillness – a death from which no life could spring, a rest from which they could never swerve until inspired with power from a source of life'.[225] Uncaused 'swerving' had long been recognised as a weak point in the Lucretian doctrine, but Jenkin thought that the atomist's conception was 'not stupid, it was simply false, as all physical explanations of the origin of energy and matter must be'.[226] While acknowledging that 'swerve' could touch upon 'free-will', he considered it impossible that natural science 'will ever lend the least assistance towards answering the Free-will and Necessity question'.[227] Nor could any insight be given into the nature of a 'First Cause'. But considerations of such questions, however important, departed from his main interest.[228]

The ancient opponents of atomic theories received short shrift from Jenkin: 'Aristotle and his followers got entangled in the "snare of words", to use Hobbes' language, and their teaching led to little or no progress in what we call science'.[229] Jenkin, anticipating modern views of the history of science, thought that it was not until the seventeenth century that serious scientific objections to the atomic theory arose, in particular in the work of Descartes and Leibniz. Descartes's ideas were ludicrous, for 'his laws of motion are false, and he knew it, but he says we must judge from our experience of gross matter; and yet this man insisted on clear conceptions as the very test of truth'.[230]

Jenkin objected to Leibniz's idea of God as 'Court of Appeal', for 'we do not use the argument now in support of circles as more perfect than other

---

[223] Colvin and Ewing (eds) 1887, II, p. 179.
[224] Colvin and Ewing (eds) 1887, II, p. 183.
[225] Colvin and Ewing (eds) 1887, II, p. 192.
[226] Colvin and Ewing (eds) 1887, II, p. 192.
[227] Colvin and Ewing (eds) 1887, II, p. 193.
[228] See the discussion in Colvin and Ewing (eds) 1887, II, pp. 193–97.
[229] Colvin and Ewing (eds) 1887, II, p. 198.
[230] Colvin and Ewing (eds) 1887, II, p. 199.

figures, and therefore more consistent with Divine wisdom'.[231] But Leibniz's monads, or ultimate units of being, while interesting and perhaps leading to Boscovich's centres of force, contrasted unfavourably with the ideas of Lucretius. The theory that atoms were voids in an ether rather than small particles with mass which moved in nothingness, was however worthy of consideration, particularly as it came from 'the man who claimed to have run a race with Newton in inventing the higher calculus of mathematics, and who enounced the doctrine of *vis viva*'.[232]

Thus in Jenkin's interpretation of the history of science, effective counters to the atomic theory were emerging even as it was being put on a strong empirical and theoretical base.[233] Yet he was satisfied that the two interpretations, continuity and discontinuity, atoms and plenum, were converging, particularly if Thomson's extensions to Helmholtz's work on vortex rings could be validated. 'Having traced the theory of a continuous fluid to its development in the hands of Thomson, we find that this school too has arrived at indestructible elastic atoms as the secondary constituents of gross matter, though they reject the crude atoms of Lucretius as a primary material'.[234]

In the 1860s it was considered untenable that solid atoms, moving in a void and acting by collision, were the fundamental and only constituents of the universe.[235] Forces such as gravitation, magnetism and electricity required either the operation of causes acting at a distance across truly empty space, or the assignment of properties to the vacuum. For electricity, magnetism, heat and light, Jenkin considered that the existence of an ether was 'almost demonstrated', and that Faraday 'by proving the influence of the intermediate material in the case of electrical action, by his discovery of magneto-optic rotation, and by showing how lines of force arose in media, rudely shook the theory of attraction and repulsion, action at a distance across a perfect void'.[236] By analogy, even gravitation might be explained.[237]

Jenkin concluded that there were three distinct atomic theories: 'atoms of 'solid singleness' ... atoms due to the motion of a continuous fluid and ... atoms having the property of exerting force at a distance'.[238] While his ontological preferences inclined towards the primacy of atoms and fields, he had to accept

[231] Colvin and Ewing (eds) 1887, II, p. 200.

[232] Colvin and Ewing (eds) 1887, II, pp. 201–2.

[233] Jenkin was thinking here of the kinetic theory of gases, and the laws of chemical combination.

[234] Colvin and Ewing (eds) 1887, II, p. 206.

[235] The 'void', Jenkin was careful to emphasise, was not only devoid of material, but without properties or immaterial entities.

[236] Colvin and Ewing (eds) 1887, II, p. 208.

[237] See the discussion in Colvin and Ewing (eds) 1887, II, pp. 207–10.

[238] Colvin and Ewing (eds) 1887, II, p. 210.

that the problem of the constitution of matter was unsolved.[239] The nascent kinetic theory of gases and the laws of chemical combination, although suggestive of atoms, were not in themselves conclusive. But as Faraday's observations had provided evidence for the existence of fields, so it could be expected that it would be possible to validate observationally the true nature of atoms.[240] It would be necessary to provide a scientific explanation, the nature of which was evident in the writings of Thomson, Clausius, Rankine and Clerk Maxwell. Then the vast difference 'between the old hazy speculations and the endeavours of modern science' would be perceived.[241]

On the face of it, deliberations on a standard of resistance would not involve profound questions of epistemology, but it is clear from various reports of the British Association committee on electrical standards that the nature of electricity required a deep consideration of philosophical problems. The recognition, definition and realisation of electrical quantities were not straightforward like those of length, time and weight; they were more akin to those of energy and force. In other words, electrical quantities were more evidently 'theory-laden'.[242] It is not certain that the philosophical undercurrents in the British Association reports on electrical standards originated with Jenkin, although he wrote many of them. His 1873 textbook *Electricity and Magnetism* shows the depth of his interest in epistemological questions. He wrote in the introduction:

> Not a single electrical fact can be correctly understood or even explained until a general view of the science has been taken and the terms employed defined. The terms which are employed imply no hypothesis, and yet the very explanation of them builds up what may be called a theory. *The terms cannot be explained by mere definitions, because they refer to phenomena with which the reader is unacquainted* [our italics].[243]

Electricity could not be understood without understanding the 'facts', which could not be understood without comprehending the theory in which the facts were embedded. Further, the facts had to be stated at the outset, for they were not immediately evident to the senses: 'Many of the assertions cannot be proved to be true, except by complex apparatus, and the action of this complex apparatus cannot be explained until the general theory has been mastered'.[244] To

---

[239] Colvin and Ewing (eds) 1887, II, p. 211.

[240] See also the discussion in J.C. Maxwell and H.C.F. Jenkin, 'On the elementary relations between electrical measurements', *Philosophical Magazine*, XXIX (1865), p. 440. This paper was a reprint from the *Reports of the British Association* for 1863.

[241] Colvin and Ewing (eds) 1887, II, pp. 213–14.

[242] Harré 1985, pp. 133, 139.

[243] p. vi.

[244] p. vii.

understand electricity the student was forced to confront problems of knowledge.

In the unfinished 'Fragment on Truth', published only after his death, Jenkin began to record his thoughts on the certainty of knowledge. Some things were more or less definable by the methods of science – the 'truth' of the 'ether', for example. Other questions, such as the free will or necessity question, could not be settled by physics or chemistry. Jenkin had come to feel that the question could not be evaded: 'It is admitted on all hands that in all matters, whether of faith, knowledge or perception, we should endeavour to attain truth; to believe truly, know truly, feel truly'. Yet so different were men's conceptions of knowledge that there was strong need 'for some criterion or touchstone ... [to] discern truth from falsehood'. The need would remain unsatisfied, for such a touchstone would not, could not, be found. The desire to have one he saw as 'rather akin to the vague longing for magic powers than to any healthy appetite'. Absolute truth, thought Jenkin, was, if not non-existent, then certainly unobtainable, except perhaps in the most trivial sense.[245]

Adopting a form of dualism, Jenkin considered that truth was a measure of 'a concordance between some verbal expression and an external fact or between some mental expression and an external fact'.[246] Mathematics gave some kind of truth in providing a sure test of success, yet was 'impotent to suggest a theory'.[247] 'Agreement between many minds as to any statement becomes more and more probable as the statement is more and more restricted to the simplest class of facts' and herein lay 'the supposed superiority of mathematics, and science generally in regard to truths'.[248] But such truths meant very little, and it was easy to agree on the 'truth of the language by which they are expressed'. Beyond the simplistic descriptions provided by mathematicians and scientists the expression of real and complex situations demonstrated that 'the statements and the facts accord imperfectly'. In practice, the lack of concordance between descriptions in words or symbols, and the real world of experience, would be obvious to 'the mechanic'. He would know the approximate nature of a 'statement and the corresponding fact' but would accept near agreements as useful. For Jenkin the 'facts' were material, or were actions on material things such as bridges and 'breaking weights', pumps and the power required to operate them.[249]

One might call this form of Jenkin's truth 'realistic dualism' – characterised by a dichotomy between the descriptions of facts and the facts themselves. His alternative definition of truth could be described as 'idealistic

[245] Colvin and Ewing (eds) 1887, II, p. 264.
[246] Colvin and Ewing (eds) 1887, II, p. 264.
[247] Colvin and Ewing (eds) 1887, II, p. 213.
[248] Colvin and Ewing (eds) 1887, II, p. 264.
[249] Colvin and Ewing (eds) 1887, II, p. 265.

dualism'. The facts were of the same type, but the accord now was 'between what seems and what is, between the conception in our minds and the fact which gives rise to that conception'.[250] He could not accept that truth was attainable by examining such concordance. The problem was one of perception, and he could see no means by which a mental impression or concept could have an identity with the object that allowed the idea to be conceived. Again he turned to a practical argument. The dent produced by a hammer is not the same as the hammer; no two coins are identical, even if struck from similar metal with the same die. So if material things, seen as analogies to mental constructs, lacked identity, how much more unlikely was it that perceptions, widely different from one person to another, could represent the actuality of objects? Idealism could not work, other than approximately. Jenkin did not object to relative or absolute truth but concluded that he had 'exposed the folly in any hope of a criterion of absolute truth'.[251]

Jenkin did pose the question of whether, if there were a criterion, it could ever be applied with exactitude. He intended to apply the question to physical measurement and mathematics, but never completed his consideration of the latter. In discussing measurement, while he recognised the existence of a personal equation, from the standpoint of his realistic dualism the mind and its sensations could be ignored. In scientific measurement, he said, 'things, not sensations, are compared'.[252] Jenkin did not distinguish between things whose measurements were obtained by 'self-measurers' or 'non-self-measurers'.[253] His 'things' included time, space, matter, energy; measurements compared the likes of position, velocity, tenacity.[254]

Jenkin shared Thomson's view of the association between knowledge and measurement, also the opinion that regardless of 'absolute truth' the numerical relationships between objects could be defined and measured scientifically. 'In these regions dispute is settled by an appeal to things not to minds: no criterion of truth is missed, and prolonged difference of opinion concerning the relations measured is impossible, except as regards a small and constantly lessening hinge or margin of the whole part'.[255] These epistemological musings were echoes of ideas originating in Jenkin's work on the electrical standards committee. In the opening section of the standards committee's second report, written in 1863, Jenkin proposed the development of a means whereby 'laws

---

[250] Colvin and Ewing (eds) 1887, II, p. 265.

[251] Colvin and Ewing (eds) 1887, II, p. 266.

[252] Colvin and Ewing (eds) 1887, II, p. 267.

[253] An example of a 'self-measurer' is a metre rule, which is itself a length and measures length. No threads of theoretical arguments are required to interpret its readings. A 'non-self-measurer' would be something like a voltmeter, whose readings are not 'volts', but numbers whose values are defined by some standard definition, and obtained, often by calibration.

[254] See Harré 1985 for a discussion of this.

[255] Colvin and Ewing (eds) 1887, II, p. 267.

remembered in their abstract form [could] be applied to estimate the forces required to effect any given practical result'. A telegraph engineer needed to be sure in advance of the limits of his devices and techniques, including exact knowledge of electrical quantities.

> All exact knowledge is founded on the comparison of one quantity with another. In many experimental researches conducted by single individuals, the absolute values of those quantities are of no importance; but whenever many persons are to act together, it is necessary that they should have a common understanding of the measures to be employed.[256]

Jenkin concluded that the 'things' forming the basis of any standard should consist solely of what he called 'primary units' – those of mass, length and time – because a system where every unit derived from primary units 'bears the stamp of the authority, not of this or that legislator or man of science, but of nature'. The 1863 report did not discuss the manner in which units could be realised, but the committee had started to consider the practical problems. Two kinds of difficulty were identified: the definition and determination of the standards themselves; and the specification and manufacture of sub-standards. Jenkin's understanding of the significance of these problem is explicitly stated in a reply to Werner Siemens on the subject of the definition and determination of resistance.[257] According to Jenkin, Siemens had not grasped the requirements of either a standard or a sub-standard.

Jenkin's metaphysics recognised the need for atoms and a plenum, and he saw the resolution of the dichotomy in terms of the ether, which could 'condense' in vortex rings to form centres of force. Thus could the world be explained. However his epistemology was empirically based: exact, perhaps true, knowledge derived from measurements associated with existent 'things', 'things' which were not hypothetical but real. Reality was assured by the physical existence of instruments and artefacts. While Jenkin held this as a general principle, it was true in particular in his electrical engineering. Nevertheless, as he wrote in the introduction to *Electricity and Magnetism*, instruments themselves in time become outdated, yet the principles upon which their construction and use depends are permanent, 'depending on no hypothesis'.

---

[256] Maxwell and Jenkin in the *Philosophical Magazine* for 1865, p. 437.
[257] Jenkin's reply to Siemens' paper, 'On the question of the unit of electrical resistance', appeared in the *Philosophical Magazine* for September 1866.

Chapter 6

Untiring Zest for Life[1]

Fleeming Jenkin's remarkable ability was not confined to engineering. Before the move to Edinburgh, he began to extend his interests beyond scientific and educational issues, into subjects which were far from the obvious domain of a professor of engineering: political economy, public health, drama and literature. On the surface these concerns seem tangential and irrelevant to the main direction of Jenkin's life, the diversions of a man who had at last achieved material success. Freed from a punishing schedule of contracting work, revelling in new intellectual circles, communicating in French, German and Italian as well as he did in English, he could at last develop interests apart from his main career.

Yet that was not Jenkin's nature. His leisure time he spent on active pursuits, reel dancing or skating, fishing, climbing, sketching, walking or sailing with his family in the Highlands.[2] Thinking and writing, in contrast, were serious, whatever the issue. Furthermore Jenkin's efforts were seriously received, his contributions to fields beyond engineering influential. He attacked his new interests with a customary verve, incisiveness and thoroughness – 'a clear-headed man who took nothing for granted, and never wrote on anything till he felt convinced that he understood it'.[3] Engineering and science were not forgotten, but submarine telegraphy, no longer offering the challenge which it had presented in 1860, receded in importance. 'It is possible', said Stevenson, 'to have too much even of submarine telegraphy and the romance of engineering'[4] – yet Jenkin's growing detachment from telegraphy stemmed not from boredom, but from his realization that many of the technological difficulties had been resolved. Later, he would return to experimental work in electrical engineering, but meanwhile there were other problems to occupy his restless intellect.

In one of his final pieces of writing – an unfinished review of a life of George Eliot – Jenkin set out his requirements of a biography:

---

[1]   Colvin 1921, p. 159.
[2]   Hole 1884, p. 110.
[3]   Tait 1888, p. 435.
[4]   RLS, p. xcv.

> A biography may reveal to us a friend or at least a companion ... The
> present book falls wholly short of any such revelation. Possibly George
> Eliot never revealed herself except by her books, yet how we long to
> know what that narrow Low Church creed meant to the girl; how we
> long to understand how she felt when the faith to which she clung left
> her wholly within a few days.... we know that this woman of pure heart
> and right moral feeling lived as wife to a man who had no right to marry
> her. How did she feel? How did passion speak to this great soul?[5]

The record of actions alone, he concludes, is no record of the true life of a man.

Much of our knowledge of Jenkin's own character, motivations and
beliefs, and hence our assumptions of why his intellectual direction shifted in
the late 1860s, derives from Stevenson's florid account. It seems that Jenkin,
although familiar with modern liberal thinking from his mother's continental
salons, had not adopted liberalism in any formal sense, and drew away from
her 'ready-made opinions' as he matured.[6] His work was never directed by
simple religious or political concerns. In religion, he was evidently a fairly
passive Anglican, although Stevenson believed that his piety 'was a thing of
chief importance'. He had written to Stevenson:

> All dogma is to me mere form; dogmas are mere blind struggles to
> express the inexpressible. I cannot conceive that any single proposition
> whatever in religion is true in the scientific sense; and yet all the while I
> think the religious view of the world is the most true view.[7]

He drew nearer to the church towards the end of his life, and during his final
year began to take communion. 'The longer I live, my dear Louis, the more
convinced I become of a direct care by God – which is reasonably impossible –
but there it is'.

There was a gradual change in character as Jenkin entered middle age.
Stevenson, who first met the professor in 1868, suggests that after his return to
Edinburgh Jenkin's mood lightened. The youthful 'inhuman narrowness' was
lost, Jenkin ceased to be 'something of the Puritan', growing ripe and mellow
and better understanding 'the mingled characters of men'.[8] This is consistent

---

[5]   'A fragment on George Eliot' was to have been a review of the biography written by
Eliot's second husband: John Walter Cross, *George Eliot's Life as Related in her Letters and
Journals, Arranged and Edited by her Husband* (Blackwood, 3 vols, 1885). The fragment was
published in Colvin and Ewing (eds) 1887, I, p.174.
[6]   RLS, p. xlvi, suggests that Jenkin had been uncritical in accepting his mother's
dogma in his youth.
[7]   RLS, p. cxxxv.
[8]   RLS, p. cxxxiv.

with Stevenson's hint that Jenkin was frustrated by the limitations of science in explaining any greater truths:

> Number and measure he believed in to the extent of their significance, but that significance, he was never weary of reminding you, was slender to the verge of nonentity. Science was true, because it told us almost nothing. With a few abstractions it could deal, and deal correctly; conveying honestly faint truths. Apply its means to any concrete fact of life, and this high dialect of the wise became a childish jargon.[9]

As a focus of Jenkin's interests, measurement was losing ground to language:

> He had a keen sense of language and its imperial influence on men; language contained all the great and sound metaphysics, he was wont to say; and a word once made and generally understood, he thought a real victory for man and reason. But he never dreamed it could be accurate, knowing that words stand symbol for the undefinable.[10]

His belief in art's central importance is spelt out in a letter to William Young Sellar in 1880. Sellar had missed the Jenkins' performance of *Agamemnon*, for which Jenkin only half jokingly admonished him: 'Do you think I will even give you credit for one grain of artistic faculty in you when you missed the chance of seeing the *Agamemnon* with such a Cassandra and Clytemnestra'. He continued:

> ... even while the tears are in my eyes I could not choose but smile to think that a Professor of Greek, the man familiar with the works of the most artistic people in the world had not one poor little word to say of the pleasures and duties of art – that to me poor mechanical mathematical creature, Art should be a great part of my existence that it should give me the most intense of all pleasures ... that from the cradle I train my boys to look for the beauty and imperative of life in Art i.e. in man's relation to nature in respect of beauty and that you had not a word to give it not a thought in your scheme of life.[11]

Of course a man may be a prophet and a poet, Jenkin continued – even an artist and an engineer. He took strong exception to Sellar's unfavourably comparing him with Tait:

---

[9] RLS p. cxxxvi.
[10] RLS, p. cxxxv.
[11] National Library of Scotland, ms 2633, ff. 208–210, Jenkin to Sellar, 21 May 1880.

What do you mean you villain by blowing up Tait to me... He voluntarily abjures some of the best parts of life – so do you – and why do you call me narrow? Religious moral political choreographic poetical artistic intellectual physical metaphysical social athletic philosophic savage don't I enjoy the lot – not very deep in any – granted – but not as narrow as you you insular Scot ...[12]

Tait's only interest outside natural philosophy appears to have been golf.[13] As to Jenkin's assessment of himself, despite the abundance of adjectives it is unduly modest about his achievements outside engineering, and also overlooks the connecting features of his many interests.

The background to Jenkin's wide range of activities lies in his upbringing and education. He had mastered three European languages besides English – German, French and Italian[14] – understood Latin and had some knowledge of classics. His mother had immersed him from an early age in music, art and literature. This artistic grounding was later reinforced by the friends of his own choosing, notably Elizabeth Gaskell, Robert Louis Stevenson and the intellectuals whom he met through the Austins. Stevenson, in an uncharacteristically critical passage where he tried to account for the tendency towards liberalism which he, Stevenson, considered illogical, placed the blame for this and other shortcomings on Jenkin's mother, 'an imperious drawing-room queen'. Thus Jenkin inherited her faults: 'generous, excessive, enthu-siastic, external; catching ideas, brandishing them when caught... ready at fifteen to correct a consul, ready at fifty to explain to any artist his own art'. And while Jenkin was meticulous in his work, 'his thoroughness was not that of the patient scholar, but of an untrained woman with fits of passionate study'.[15] Yet Jenkin was far from being so dogmatic or so ephemeral in his writings and interests, and Stevenson here seriously understates the weight, erudition and consistency of Jenkin's intellectual work.

The common threads uniting Jenkin's early and later interests, both artistic and scientific, theoretical and practical, were language and communication. Mathematics and measurement were treated by him as a means of transmitting information; many of his other interests were similarly focused. So, for instance, work on the sound and articulation of the human voice grew out of his interest in electrical science but was also related to a fascination with

[12] National Library of Scotland, ms 2633, ff. 208–210, Jenkin to Sellar, 21 May 1880. Punctuation is mainly absent from the original.
[13] Tait's golf clubs and phosphorescent golf balls, which he developed for playing in the dark, are now on display in the scientific instruments gallery of the Royal Museum of Scotland. For this fascinating piece of information we are grateful to Alison Morrison-Low.
[14] Colvin 1921, p.154.
[15] RLS, p. xlvi.

mechanisms of theatrical communication.[16] Jenkin found that, although not a great writer, he had the born teacher's talent for communicating difficult subjects to a general audience. Unlike Thomson, he had the desire and the ability to reach non-specialist readers. This is shown in the way that the two wrote on Charles Darwin's theory of evolution.

> Jenkin had expressed the thermodynamicists' thesis with unsurpassed clarity, bringing it alive for his readers in a way that Thomson never accomplished. Where Thomson was arrogant, Jenkin was humorous. While Thomson drew his arguments from Fourier analysis with differential equations, Jenkin drew from rainfall and ballgames. The lyricism of Jenkin's writing stands in marked contrast to the wooden prose of Thomson. Jenkin's review is distinguished by its appeal to the educated but technically untrained public – the rapidly expanding audience for the Victorian reviews.[17]

The Darwin critique appeared in the *North British Review*, which had published Jenkin's first work aimed at a wider public, on submarine telegraphy in 1866. The success of this first article appears to have encouraged him to write on more diverse topics. His enquiring mind and broad educational background meant that although Jenkin remained essentially a critic in some of these other disciplines, he became a respected and influential one. As his friend Sidney Colvin wrote after Jenkin's death, 'on whatever subject engaged his attention he was almost sure to find something to say that was well worth hearing. When masters in such divers fields as Darwin and Munro have acknowledged the value of his arguments, weaker testimony is needless'.[18] Alfred Ewing, looking back upon the period when he was closest to the Jenkins, described the pleasure of associating with the professor and his gifted wife – for many of Jenkin's interests were shared ones with Anne:

> He had too many facets to confine himself to engineering. His writings on other subjects – economics, literature, the drama, Greek dress, English rhythms, the atomic theory, natural selection – are evidence not only of his variety but of his insight, of his ability to throw fresh light on anything he took up.[19]

Ewing observed that Jenkin's enthusiasm, his joy in living, was reflected in his talk, 'which was always ready and forceful and often witty'. Jenkin was developing a reputation as a great talker. The dramatist Brander Matthews

---

[16] Morris 1994, p. 321.
[17] Morris 1994, p. 335.
[18] Sidney Colvin, 'Prefatory Note to Jenkin's collected papers' in Colvin and Ewing (eds) 1887, p. clxxiii.
[19] Ewing 1933, p. 251.

recalled that the first time he met Jenkin he felt 'surprised delight in his scintillating discourse' and admired 'the marvellous range of his interests and of his attainments.[20] Colvin perhaps found the key to Jenkin's character:

> In conversation and human intercourse lay perhaps his chief pleasure of all. His manly and loyal nature was at all times equally ready with a knock-down argument and a tear of sympathy. Chivalrous and tender-hearted in the extreme in all the real relations and probing circumstances of life, he was too free himself from small or morbid susceptibilities to be very sparing of them in others, and to those who met and talked with him for the first time might easily seem too trenchant in reply and too pertinacious in discussion.[21]

He did not always use his talent for conversation seriously, as the wife of an Edinburgh colleague recollected:

> He told me once that he shocked some of his colleagues. He would take up a subject – say the desirability of Polygamy – and defend it with endless ingenuity and wit. Most of the other professors would either sit in pained silence, or proceed to remonstrate with him ...'[22]

There are echoes in this of Jenkin's comment about the benefits for engineers, both professors and students alike, of life in a university, of the discourse with non-scientists: 'I think that I myself am better for the professors that I meet with and associate with'.[23] Talking was his *raison d'être*.

> More even than his writing, Jenkin's talk displayed the variety of his interests and the readiness with which the flash of his restless intelligence could be turned to lighten up any subject, however far from his usual thoughts. It was impossible to take him unawares. Ready on the moment to produce a theory of things in general or in particular, he would lay it down with a dogmatism which was half jesting, half serious, but wholly good natured and full of wit and point. His versatility would have been dangerous had it not been tempered by strong common-sense, and by his power of giving concentrated application to any matter he took seriously in hand.[24]

---

[20]   Matthews (ed.) 1958, pp. 72–3.
[21]   Colvin 1921, pp. 157–58.
[22]   Jenkin mss, Mrs Roscoe's recollections of Anne Jenkin.
[23]   *Devonshire commission*, p. 103. See Chapter 5, above.
[24]   Obituary, *Institution of Civil Engineers Proceedings*, p. 369.

His fluency and scientific standing made him in demand as an expert witness in Scottish courts, and it was said that no-one could put a better case before a jury.

> But Jenkin was too honest to be a good partizan unless his conviction lay on the side he supported; delighted as he often was to support a paradox as an exercise in dialectics, he would never, where serious issues were at stake, add his weight to the side he did not believe in.[25]

Stevenson described Jenkin, under the pseudonym Cockshot, in his essay 'Talk and Talkers' as 'vastly entertaining':

> [He] has been meat and drink to me for many a long evening. His manner is dry, brisk, and pertinacious, and the choice of words not much. The point about him is his extraordinary readiness and spirit... He is possessed by a demoniac energy, welding the elements for his life, and bending ideas, as an athlete bends a horseshoe, with a visible and lively effort.[26]

Jenkin was an early member of the Savile Club, founded in 1868 as the New Club, where he stayed when visiting London. The club was in Spring Gardens, near his old Duke Street office, and moved to Savile Row in 1871.[27] One attraction of membership was the quality and price of the cuisine, as noted by Stevenson, who joined the Savile in 1874:

> I like my club very much; the *table d'hôte* is very good: it costs three bob: Two soups, two fish, two entrees, two joints and two puddings; so it is not dear; and one meets agreeable people.[28]

The policy of the Savile was informality, 'founded on a principle aimed against the standoffishness customary in English club life'.[29] Many members were drawn from artistic or literary occupations, such as Sidney Colvin, but there were also scientists – Darwin was a member.[30] Jenkin introduced a

---

[25] Obituary, *Institution of Civil Engineers Proceedings*, p. 374.

[26] Stevenson 1911, pp. 89–90.

[27] Information about the history of the Savile Club has been kindly provided by its secretary, Nicholas Storey, and includes extracts from C.E.S. Phillips, *The Savile Club, 1868–1923*, (privately printed). Jenkin was a committee member from 1882 until 1884.

[28] Booth and Mehew (eds) 1994, II, p.27, Stevenson to his mother, 3 July 1874. Stevenson had been proposed and seconded for membership by Colvin and Jenkin.

[29] Colvin 1921, p. 119.

[30] It is not clear how well Darwin and Jenkin knew each other. Jenkin and Charles Darwin's son Horace were acquainted in the 1880s, taking out a joint patent on water motors in 1884. See M.J.G. Cattermole, *Horace Darwin's Shop: a History of the Cambridge Scientific Instrument Company, 1878–1968* (Bristol: Adam Hilger, 1987), for the reference to which thanks to Anne Locker of the IEE Archives.

number of his friends to membership of the club, among them Charles Hockin, civil engineer and fellow of St John's, Cambridge, the Edinburgh publisher David Douglas, and Thomson and Stevenson.[31] At the 'long table' members were expected to be ready to talk and 'liable to accost without introduction'.[32] It was by this means that Brander Matthews encountered Jenkin for the first time.

> I met him in the dingy smoking room of the Savile Club in 1881. We had fallen into talk over coffee, without any formal introduction, as is the kindly custom of the Savilians; and I remember the eagerness with which I slipped into the adjoining room to inquire from Andrew Lang the name of the brilliant conversationalist with whom I had unwittingly forgathered.[33]

At first Jenkin had not been popular in his club, being known only as 'the man who dines here and goes up to Scotland'.[34] His 'porcupine ways', according to Stevenson, had tended to thwart friendship. Frewen Jenkin later wrote that he could never understand 'why Papa seemed so fond of contradicting everyone and of saying things exactly as he meant them, especially to people we wished to make friends with'.[35] Yet from his berth at the Savile Club, Jenkin eventually came to enjoy 'a late sunshine of popularity'. He wrote to Stevenson:

> Will you kindly explain what has happened to me? All my life I have talked a good deal, with the almost unfailing result of making people sick of the sound of my tongue. It appeared to me that I had various things to say, and I had no malevolent feelings, but nevertheless the result was that expressed above. Well, lately some change has happened. If I talk to a person one day, they must have me the next. Faces light up when they see me. – 'Ah, I say, come here' – 'come and dine with me.' It's the most preposterous thing I ever experienced. It is curiously pleasant. You have enjoyed it all your life, and therefore cannot conceive how bewildering a burst of it is for the first time at forty-nine.[36]

By the time of his death, Jenkin was even spoken of as 'the best talker in London'.[37]

The circle of friends was growing. When living at Claygate and working in London, Jenkin had been largely absorbed by scientific and engineering

[31] Information from Nicholas Storey.
[32] Colvin 1921, p. 119.
[33] Matthews (ed.) 1958, p. 72.
[34] RLS, p. cxli.
[35] Yale, Beinecke, 4982, C.F. Jenkin to Stevenson, 9 August 1887.
[36] RLS, p. cxli.
[37] Obituary, *Institution of Civil Engineers Proceedings*, p. 369.

work. In Edinburgh he found new acquaintances, and rediscovered old ones, from many walks of life. The change in character upon which Stevenson remarked, the mellowing of the porcupine ways, coincided with the widening of his social circle.

An outlet for Jenkin's popular journalistic work was provided by one new friend, David Douglas, editor of the *North British Review*, to whom he had been introduced by Peter Tait.[38] The article on submarine telegraphy was followed by a critique of Darwin and a review of a work on population statistics, both in 1867, then a discussion of the legitimacy of trade unions and the review article on Munro's *Lucretius*, which appeared together in the first issue of 1868. Douglas's editorship terminated the following year when the journal was taken over by a group of liberal Catholics, and Jenkin wrote for it no more.[39] The opportunity given him by Douglas to write in a popular style, and on new subjects, had presented Jenkin with a vent for his outspokenness. Journalism also brought some much needed income in the months before Jenkin's move to Edinburgh.[40] The *North British* published articles anonymously, and was the scene of robust debate. Jenkin revelled in this, writing to Douglas: 'I hope for hearty abuse, which is much better than to be passed over in silence. I am confident I can crack my opponents' heads if they will only fight'.[41] At the time Jenkin was awaiting confirmation of the Edinburgh chair, and asked Douglas to respect his anonymity: 'In about three months' time I shall not care who knows it, but till then it had better be kept strictly incognito'.[42] But far from attracting insults, the 1867 and 1868 articles left a lasting impression in their various fields. Darwin and Munro each accepted some of Jenkin's arguments, the trade union paper contributed to John Stuart Mill's reconsidering the wages-fund concept, and Dr Matthews Duncan, author of *Fecundity, Fertility and Sterility*, paid the rare compliment of reproducing Jenkin's review as an *addendum* to his second edition.[43]

The subject matter of the *North British* articles reflects a deep-rooted concern with issues apart from science and engineering. In fact Stevenson's suggestion of 'inhuman narrowness' in the younger Jenkin is misleading, for

---

[38] Morris 1994, p. 322.

[39] Houghton (ed.) 1966, pp. 663–5. The *North British Review* had a circulation of about 3000 at this time.

[40] Jenkin received £40 from Douglas in March 1868, which may have been in payment for both the Trades Unions and Lucretius articles: National Library of Scotland, Douglas papers, letter 114, Jenkin to Douglas, 2 March 1868. The Douglas papers are quoted by kind permission of the Douglas family.

[41] National Library of Scotland, Douglas papers, letter 114, Jenkin to Douglas, 2 March 1868.

[42] National Library of Scotland, Douglas papers, letter 113, Jenkin to Douglas, 12 February 1868.

[43] Colvin 1921, p. 156; obituary, *Institution of Civil Engineers Proceedings*, p. 369.

the love of literature and drama nurtured by his mother from his earliest years had come to be a passion. With Shakespeare he had soothed himself on telegraph voyages.[44] He was widely read in German, French and Italian literature, and especially knowledgeable about French drama.[45] Recognised in his later years as an authority on Greek plays and English rhythms, it was said that Jenkin knew 'more of the construction of plays than any man in England.'[46]

Jenkin's only attempt to use this theoretical understanding in practice was the play *Griselda*, which he wrote in 1881.[47] It is revealing about him that writing was not stimulated by poetic feelings; his intention was to redress an unbalanced treatment of one character in earlier versions of the story by Boccaccio and Chaucer.[48] The engineer-poet had another purpose: 'to treat a story ... like a sum in arithmetic' – his own explanation.[49] As drama, *Griselda* was not a success. Stevenson, who wrote in his memoir of Jenkin that the play did not work, was disingenuous in telling Jenkin that *Griselda* was 'most excellent ... it is written stunning'.[50] He did not believe that Jenkin generally wrote well, conceding only that he 'sometimes wrote brilliantly, as the worst of whistlers may sometimes stumble on a perfect intonation'.[51] A mainly enthusiastic review by William Archer of the Jenkin memoir challenged Stevenson for underrating 'the literary faculty of his friend' and went on to say that Jenkin 'was no mere amateur in the use of his pen'.[52] Not so, said Stevenson in a private reply to Archer. While Jenkin was often amusing, 'I never, or almost never, saw two pages of his work that I could not have put in one without the smallest loss of material. That is the only test I know of writing'.[53] Jenkin had teased his friend that literature was no trade or craft, that 'the professed author was merely an amateur with a doorplate', to which Stevenson responded that writing was 'as much a trade as bricklaying, and you do not know it'.[54] Stevenson, along with the poet W.E. Henley, tried to introduce him to this trade, but thought that he had succeeded only in developing Jenkin's critical faculties.

---

[44] RLS, p. xciii.

[45] Hole 1884, p. 110.

[46] Obituary, *Institution of Civil Engineers Proceedings*, p. 369.

[47] The play was privately printed and remained unpublished during his lifetime. Its only known performance was in the Jenkins' private theatre in January 1882. It appears in the collected papers: Colvin and Ewing (eds) 1887, I.

[48] The story of Griselda is the final one in Boccaccio's *Decameron*, and is the Clerk's Tale in Chaucer's *Canterbury Tales*.

[49] RLS, pp. cxxiv–cxxv.

[50] Booth and Mehew (eds) 1994, III, p. 189, Stevenson to Jenkin, June 1881.

[51] RLS, p. cxxxix.

[52] Archer 1888.

[53] Booth and Mehew (eds) 1995, VI, p.113, Stevenson to Archer, *c*.12 February 1888.

[54] RLS, p. cxxxix.

So although Jenkin appreciated and enthused about great art, especially drama, he discovered that he could not himself create it. Nor did he have great talent as an actor. There were, though, other ways to satisfy his lifelong love for 'the play and all that belonged to it'.[55] He was an engineer, and so he applied himself to understanding the construction of plays, of poetry, even of theatrical costume. He also became a respected critic, and a producer and stage manager of plays in the Jenkins' own private theatre. These performances started in 1870 in their first Edinburgh home, 5 Fettes Row. When the family moved to a large house in fashionable Great Stuart Street in 1873, Jenkin was able to arrange there a small theatre, making a stage by letting down a dining room wall into the boys' playroom behind, with the whole of the dining room as auditorium. Usually two plays were staged each spring, with five performances of each, some to servants and others to privately invited audiences of friends.[56] Occasionally the company played elsewhere – the *Agamemnon* was performed in London in 1880 to great acclaim, and the 1877 programme was repeated later in the year at St Andrews Town Hall.[57] Performances followed weeks of rehearsals, with Jenkin 'an iron taskmaster'.[58] He selected and adapted the plays, designed and arranged costumes and accessories, applying himself to even the smallest detail: 'I have acquired the art of beard-making from an ancient Jew ... I am a very promising pupil of his'.[59]

Robert Louis Stevenson, the engraver William Hole, and Professor Lewis Campbell of St Andrews, an old friend from the Edinburgh Academy, were, along with Ewing, regular members of the company. Anne Jenkin was the star – 'As for Mrs Jenkin, it was for her that the rest of us existed and were forgiven', wrote Stevenson[60] – and her husband delighted in her success. He told Ewing that she had 'surpassed herself' as Cassandra and Clytemnestra in the *Agamemnon*.[61] He was not alone in admiring her ability. Sidney Colvin described her performances as an unforgettable experience. 'Her features were not beautiful, but had a signal range and thrilling power of expression'. He agreed that Clytemnestra was her greatest role: 'I can vouch for having seen on no stage anything of greater – on the English stage nothing of equal – power and distinction'.[62] Ewing felt that she could have achieved greatness as a professional actress. Anne Jenkin's talent had, according to Stevenson, been inherited from her maternal ancestors, the Barrons of Norwich.[63] She was

55 RLS, p. cxxiv.
56 Ewing 1933, pp. 263–64.
57 Ewing1933, pp. 268, 272.
58 RLS, p. cxxviii.
59 Ewing 1933, p. 272.
60 RLS, p. cxxvi.
61 Ewing 1933, p. 272.
62 Colvin 1921, pp. 160–61.
63 RLS, p. cxxvi.

encouraged in it from an early age, the Austins staging amateur dramatics which had sometimes included Mrs Gaskell's daughters.[64] Anne Jenkin's mother, *née* Eliza Barron, also acted, joining in the Edinburgh theatricals where her 'refined dignity showed to advantage in various elderly parts'.[65]

The Jenkin theatre usually presented Shakespeare, light classics in translation, or melodrama. Stevenson joined the company as a prompter in 1871, and had a small part in *The Taming of the Shrew* the following year.[66] In 1873 the first Greek play was presented, *The Frogs*, in Hookham Frere's translation, alongside *My Son-in-Law*, a translation of Emile Augier's comedy, *Le Gendre de Monsieur Poirier*, in which Stevenson played Vatel.[67] Stevenson was the Duke Orsino in *Twelfth Night* in 1875: 'I am not altogether satisfied that I shall do Orsino *comme il faut*, but the Jenkins are pleased, and that is the great affair'.[68] In mid-May 1877, the author was 'down to the waist in Jenkin's theatricals, where I play one leading part and one subordinate'.[69] A photograph survives of Stevenson as Sir Charles Pomander in *Art and Nature*, an adaptation of *Masks and Faces* by Charles Reade and Tom Taylor; he was also to play the messenger in *Dejanira*, Lewis Campbell's translation of the *Trachiniae* of Sophocles. Stevenson, in character, appears in Figure 6.1.[70] This was Ewing's first experience of Jenkin's theatre, in the 'invisible roles of call-boy and property man'.[71] The repeated performances of the 1877 plays at St Andrews marked Stevenson's last appearance as an actor.

There was no theatre in 1878, but the plays resumed in 1879 with *Antony and Cleopatra*. Lewis Campbell, 'who had never acted in his life', played the lead opposite Mrs Jenkin. The following year came the acclaimed *Agamemnon*, taken to London in the autumn; *Griselda* in 1882, abridged versions of the *Andromache* and the *Merry Wives of Windsor* in 1883, and finally in 1884 De Musset's *May Night and October Night*, scenes from *The Rivals* and a revival of *La Joie fait Peur*, with Jenkin himself taking the part of an old servant.[72]

Fleeming Jenkin's experiences during his youth in revolutionary Europe may have influenced his taste for melodrama.[73] More likely the passion for all things theatrical had origins in his early love of literature, taking root in his

[64] Chapple and Pollard (eds) 1966, p. 831, Gaskell to Marianne Gaskell, 10 March 1851.
[65] Ewing 1933, p. 265.
[66] Booth and Mehew (eds) 1994, I, p. 222 n. 1.
[67] Booth and Mehew (eds) 1994, I, p. 432, Stevenson to his mother, January 1874.
[68] Booth and Mehew (eds) 1994, II, p. 115, Stevenson to Frances Sitwell, 5 February 1875.
[69] Booth and Mehew (eds) 1994, II, p. 209, Stevenson to Colvin, mid-May 1877.
[70] The photograph is reproduced in this volume, from an original in The Writers' Museum, Edinburgh.
[71] Ewing 1933, p. 266.
[72] Ewing 1933, pp. 264–73.
[73] Morris 1994, p. 321

wife's considerable acting talent and his own inclinations as an engineer to construct and organize. The theatre also allowed Jenkin to work closely with literary friends whose work fascinated him. In particular, his intimacy with Stevenson was based upon a shared enthusiasm for literature and drama. Sometimes this was structural rather than strictly artistic, for instance in their discussions of scansion – the engineer was disposed to consider form and detail even in the written word.[74] As a literary critic, Jenkin attracted respect but did not necessarily persuade others to his views. His series of articles about rhythm in English verse, in a review of Edwin Guest's *History of English Rhythms*, were described by William Archer as 'ingenious but unconvincing'.[75] His friend Sidney Colvin was also less than fulsome in summarising Jenkin's ideas on literature and drama as 'well worth hearing, though often one-sided and dogmatic'.[76]

Nonetheless there was merit in Jenkin's writings on drama. They were not mere dry dissection, for he held a play's emotional appeal as being of central importance. 'If I do not cry at the play, I want my money back'.[77] Yet his scientific training and analytical approach achieved a 'comprehensive understanding of the basic laws of acting' which Brander Matthews believed was rare among professionals and unheard of in an amateur.[78] From this Jenkin produced a series of papers which 'disentangl[ed] the principles of acting'.[79] Two of these described performances by Sarah Siddons, who had retired from acting in 1812 and died before Jenkin was born. He based the articles upon detailed notes and textual annotations made while Siddons was on stage, by Professor George Joseph Bell (1770–1843), father of the Manchester maths tutor. Jenkin's theme in introducing 'Mrs Siddons as Lady Macbeth' was the transience of performance art.

> We leave the theatre and think, 'Well, this great thing has been, and all that is now left of it is the feeble print upon my brain, the little thrill which memory will send along my nerves, mine and my neighbours'; as we live longer, the print and thrill must grow feebler, and when we pass away the impress of the great artist will vanish from the world.[80]

---

[74] Booth and Mehew (eds) 1994, III, pp. 189–90, Stevenson to Jenkin, June 1881. Jan Hewitt of the University of Teesside is responsible for this insight.

[75] Archer 1888.

[76] Colvin 1921, pp. 156–57.

[77] Matthews (ed.) 1958, pp. 73–74, quoting RLS.

[78] Matthews (ed.) 1958, p. 71.

[79] Matthews (ed.) 1958, p. 73. Jenkin included the paper on Mrs Siddons as Lady Macbeth among his most significant non-scientific publications:, Cambridge, J31, Jenkin to Thomson, 7 February 1883.

[80] Colvin and Ewing (eds) 1887, I, p. 45.

This was a commonplace among Victorians, that an actor's work would die with him. Fleeming Jenkin tried to provide at least a partial remedy, in publishing Bell's record of Siddons' appearance and gestures, the sound of her voice, its rise and fall and her use of pauses, along with Jenkin's own interpretation of the performance. The 'Siddons' articles, along with another paper on acting, a review of *Talma on the Actor's Art* with a preface by Henry Irving, appeared from the late 1870s. Simultaneously Jenkin was becoming fascinated by an innovation which would be a first step towards preserving the work of an actor in perpetuity – the phonograph.

Jenkin's other literary work included a review of translations of the *Agamemnon* and *Trachiniae* of which he was particularly proud, arguing that 'these old Greek plays were true dramas and could be acted even now with success'.[81] He was also responsible for a paper 'On the Antique Greek Dress for Women', a subject so esoteric that it is hard to see where it fits into the *oeuvre* of a professor of engineering. His approach to this apparently trivial and tangential issue illustrates how he dealt with all problems, through thorough investigation and a desire to share his solution with a wider public. The first Greek play performed by his company had been fitted out by a professional costumier, with, as Stevenson remembered, 'unforgettable results of comicality and indecorum'.[82] Jenkin decided in future to arrange Greek costumes himself, and embarked upon a reading of antiquarian texts, which he found contradictory and unhelpful. It took only one visit to the bas-reliefs and statues in the British Museum, where his friend Colvin was curator, for Jenkin to develop a theory of Greek tailoring. He then designed the costumes and oversaw their making. His article, published in the *Art Journal* in 1874 and illustrated with his own drawings, explains how to make and wear ancient Greek women's clothing. He made no special claim for this work, only that he had taken some pains to ascertain the facts, and was glad to have shown that the Greeks were masters of tailoring as well as of architecture, statuary and drama. Jenkin had solved the problem by reverse engineering – deconstruction and reconstruction – and the result was a small contribution to art.

Jenkin's greatest gift to literature may have been less direct, through his support of and influence upon Robert Louis Stevenson. The two had met when Stevenson was eighteen and suffering acutely a 'ferment of youth' relating to his troubled relationship with his father.[83] Jenkin, seventeen years older, though occasionally stern was never authoritarian like Thomas Stevenson. The Jenkins were hardly parent substitutes to the young Stevenson, rather offering him a glimpse into another world, liberal and intellectual, their home, according to

---

[81] Kelvin, Cambridge, J31, Jenkin to Thomson, 7 February 1883.
[82] RLS, p. cxxvii.
[83] Booth and Mehew (eds) 1994, I, p. 211.

Ewing, 'a haven, an oasis in a desert of convention and prejudice'.[84] The attraction between the Jenkins and Stevenson was mutual. Stevenson and Anne Jenkin met soon after the Jenkins' arrival in Edinburgh, and she announced to her husband: 'I have made the acquaintance of a Poet!'. 'From that day forward', she wrote, 'we saw him constantly ... our affection and admiration for him, and our delight in his company, grew.'[85] The author was as close to Anne Jenkin as to Fleeming, and a recent Stevenson biographer names her as the first of four older women with whom he forged close emotional bonds.[86] Alfred Ewing, five years younger than the author, gleaned from later conversations with Stevenson that Jenkin had been a strong moral influence.

> For Stevenson in his turbulent youth, questioning everything and impatient of authority, nothing could have been more salutary than to find so lofty a standard of conduct, so clear and simple a philosophy of morals, in a man who was no puritan, who loved and understood him, who cared intensely for the things for which he cared, and whose zest in life was equal to his own.[87]

In a letter of condolence after Jenkin's death, Stevenson wrote to Anne Jenkin: 'You know how much and for how long I have loved, respected and admired him ... I never knew a better man nor one to me more lovable'.[88] Knowing Jenkin was 'a great store of pleasure'.[89] Alfred Ewing thought that Stevenson owed to them

> much happiness, and other things perhaps more important than happiness: it was a liberal education for any young man to associate with Jenkin and his gifted wife, an enriching experience, a sharpening of even the sharpest wits, a training of mind and taste, of manners and morals. The dullest visitor to the house must have been conscious of its atmosphere of distinction – intellectual, aesthetic, ethical. Some may have found the atmosphere too rare for comfortable breathing: but for Louis it was the breath of his nostrils.[90]

Stevenson, who generally avoided polite society in Edinburgh, found continuing pleasure in the Jenkins and their circle, and in their private theatre.

---

[84] Ewing 1933, p. 250.
[85] *The Edinburgh Academy Chronicle*, March 1895.
[86] McLynn 1993, p. 38.
[87] Ewing 1933, p. 262.
[88] Booth and Mehew (eds) 1995, V, pp. 114–15, Stevenson to Anne Jenkin, 14 or 15 June 1885.
[89] Booth and Mehew (eds) 1995, V, p. 132, Stevenson to his mother, mid October 1885.
[90] Ewing 1933, p. 250.

He consulted Jenkin for advice when he contemplated marriage; the response was curt: 'A man should marry when he wants to and should not when he does not want to'.[91] Near the end of his life, Stevenson acknowledged that to Colvin and Jenkin he owed his safety – the word he used – during his most difficult times.[92]

Figure 6.1: Robert Louis Stevenson appearing as Sir Charles Pomander in Jenkin's production of *Art and Nature*, 1877 (Copyright: The Writers' Museum, Edinburgh)

---

[91]   Yale, Beinecke, 4987, Jenkin to Stevenson, 1 February 1880.
[92]   Booth and Mehew (eds) 1994, I, p. 44.

How much Jenkin directly influenced Stevenson's literary work is less clear. He tried to steer Stevenson away from writing plays, and particularly warned against continuing a collaboration with Henley. Jenkin was not alone in thinking that drama did not suit Stevenson, and although recognising that Stevenson was trying to assist their mutual friend Henley, who saw the theatre as a way to financial success, both he and Anne were critical of the results.[93] The first jointly written play, *Deacon Brodie*, was performed only once in London. After reading *Hester Noble*, Jenkin urged Stevenson to give up trying to write for the stage: 'I am not sure that Henley could not write a play but if so you are hindering not helping him'.[94] Anne Jenkin, universally known as Madam to friends and family, joined her husband in criticising the pair's next efforts, *Beau Austin* and *Admiral Guinea*. Henley wrote to Stevenson: 'Madam has read the *Beau*, it appears, and the *Admiral* both; and has determined that neither will act ... the gist of her complaint appears to be that we can neither feel our personages nor write a single sentence that can be said.'[95] The Jenkins were too expert drama critics to be easily dismissed, as Stevenson himself recognized: '[Jenkin] was one of the not very numerous people who can read a play: a knack, the fruit of much knowledge and some imagination, comparable to that of reading score'.[96] Their judgement on this proved sound, for the plays never enjoyed the success of Stevenson's other work, and the collaboration ultimately helped undermine his relationship with Henley.

As the arts had been a part of Jenkin's life from childhood, so also did the concern with political economy originate in his youth. Exposure to political events and political thinking in France and Italy generated the interest, and it was brought into focus through his contact with engineering artisans as a young man, his sympathy with both trade unionists and non-society men, and through his own business dealings during the 1860s. Jenkin's father-in-law, Alfred Austin, described by him as 'one of the cleverest men in London', was a strong influence in sharpening his understanding of political economy.[97] Austin (1806–84) was the youngest and least known of three sons of Jonathan Austin of Creeting Mill, Suffolk, who had made a fortune from government contracts during the Napoleonic Wars. The eldest son, John Austin (1790–1859), a lawyer turned academic, at one time held the chair of jurisprudence at University College, London and had written what Jenkin considered 'perhaps

---

[93] See Booth and Mehew (eds) 1994, I, pp. 54–62, on Stevenson's relationship with Henley; Matthews (ed.) 1958, p. 73, on Stevenson's misreading of the construction of plays.

[94] Yale, Beinecke, 4988, Jenkin to Stevenson, 15 June 1880.

[95] Booth and Mehew (eds) 1995, V, p. 61, Stevenson to Henley, 2 January 1885, and footnote.

[96] RLS, p. cxxiv.

[97] National Library of Scotland, Douglas papers, letter 112, Jenkin to Douglas, 10 February 1868.

the greatest English work on jurisprudence'.[98] Charles Austin (1799–1874), the second of these brothers, was a phenomenally successful barrister who earned unprecedented sums of money representing railway interests during the railway mania – estimates of his income for 1847 alone range from £40,000 to £100,000.[99] Alfred's brothers were intimates of John Stuart Mill. Charles, an ardent Benthamite from a time when utilitarianism was still a novelty, was Mill's contemporary at Cambridge. John Austin was a close friend and neighbour in Westminster of Jeremy Bentham and of John Stuart Mill's father James, and his wide circle of acquaintance also included Thomas Carlyle. For a time, John Austin lived in Paris until, like the Jenkins, he was driven away by the insurrection of 1848. Unlike the Jenkins, the first-hand experience of revolution caused him to abandon radical views, and in his old age John Austin wrote a pamphlet attacking any extension of the vote to working men. By the time Jenkin knew the Austins, Alfred's brothers had ceased to be active radicals; no evidence has been found that Jenkin actually met Mill through the Austins, although he was certainly very familiar with Mill's work. Alfred Austin's own writings and public pronouncements are those of a civil servant,[100] but he helped develop his son-in-law's more emphatically expressed views on social issues. A congruence between the interests of the two can be seen in Jenkin's article on population in 1867, the acclaimed review of Duncan's *Fecundity, Fertility and Sterility*.[101]

This article – it was actually much more than a review, raising questions and suggestions significant enough for Duncan to include the whole of it as appendix to his second edition – is not usually listed as one of Jenkin's economic papers.[102] It was rather a precursor to his work in economics, for it considered how statistical and mathematical techniques could be applied in order to explain and influence social phenomena. Another thread is discernible,

[98] National Library of Scotland, Douglas papers, letter 112, Jenkin to Douglas, 10 February 1868.
[99] See the *Dictionary of National Biography* for both John and Charles Austin; also R.W. Kostal, *Law and English Railway Capitalism, 1825–1875* (Oxford: Clarendon, 1994), p.123, on Charles Austin's career as a Q.C.
[100] Alfred Austin gave evidence to a Select Committee on the Glasgow Poor Rate Bill in 1840, and his published work includes his report as an assistant Poor Law Commissioner on women and children working in agriculture in 1843, and a report on cases of controverted elections in 1844.
[101] This appeared under the title 'Population' in the *North British Review*, XLVII OS, VIII NS (December 1867), pp. 441–62.
[102] Summaries of Jenkin's contribution to the development of economic thought can be found in Collison Black 1987, pp. 1007–8, and in Brownlie and Lloyd Prichard 1963, pp. 204–16. Five papers on political economy – the three published in 1868, 1870 and 1871–2, along with two unpublished manuscripts – were included by Colvin and Ewing in the collected works, and later reprinted as No. 9 in *London School of Economics Series of Reprints of Scarce Tracts in Economics and Political Science*, 1931, alongside works by such distinguished economists as Alfred Marshall and Nassau Senior.

a link to his longstanding interest in the work of Darwin. Both these reviews, of Darwin and Duncan, appeared in the *North British Review* for 1867, although the Darwin had been written in 1863 or earlier. Jenkin himself placed *Fecundity* among a half dozen of his most significant non-scientific papers.[103]

Despite the title, *Fecundity, Fertility and Sterility* was a work of medical statistics, an analysis of a survey taken in 1855 of childbirth in Edinburgh and Glasgow. Jenkin praised Duncan for his objectivity and thoroughness, for extrapolating from limited data to suggest fertility rates for women of various ages. He also used the review to develop an argument about the lamentable general ignorance surrounding conception and childbirth. Jenkin argued for further data to be collected, particularly to determine the risks of mothers dying in childbirth, so that couples could make informed decisions about conceiving further children. He thought that the survey could usefully be extended to produce information relating family size to children's health. Couched in the most delicate language, his paper contained the controversial suggestion of a more open discussion around the issue of limiting family size in order to improve mothers' and children's health and reduce poverty. He knew full well that determining a desirable level of population, in relation to available resources, was a complex and contentious matter. In moving beyond a purely statistical discussion and applying lessons from his own observations, Jenkin ventured into a moral sphere, further than Duncan and other statisticians.

The *Fecundity* article in many ways epitomises Jenkin's work in the social sciences. It alights on an issue which was topical and controversial, and contrives to make it more so by challenging orthodoxies and by introducing contentious themes, in this case from Darwin. As ever, he sought a robust debate: 'I hope for hearty abuse ...' The method employed was also typical Jenkin, applying mathematical or statistical expertise to social issues, lacing it with references to his own experiences, the whole presented in a form accessible to the reading public.

The first of Jenkin's three major papers on economics appeared the following spring.[104] 'Trade Unions: How Far Legitimate?' latched on to another current theme, for a Royal Commission on trade unions was then in progress. Jenkin had been provoked to respond to Robert Lowe, who in supporting the wages fund doctrine in the *Quarterly Review* had concluded that trade unions were incapable of delivering material benefits to their members.[105] Jenkin saw this theoretical problem as fundamental to settling public policy: if it were indeed impossible for unions to raise wages or otherwise improve working conditions, then any debate about whether they should be allowed to try was

---

[103] Kelvin, Cambridge, J31, Jenkin to Thomson, 7 February 1883.
[104] In the *North British Review*, XLVIII OS, IX NS (March 1868), pp. 1–62.
[105] Robert Lowe, 'Trades Unions', *Quarterly Review*, CXXIII, October 1867, pp. 351–83. Lowe (1811–92) was 1st Viscount Sherbrooke.

pointless. He started by analysing the underlying mechanisms, rather as he would have approached a technical or scientific project. Reasoning his way to a conclusion that the wages fund concept was invalid, he moved from theory on to concrete examples from his observations of industry, investors and consumers. The wages fund idea saw manual labour as homogeneous. Jenkin, unusual among his class in his knowledge of shopfloor relations, shared with his readers the fruits of years of observing workshop life: the differing degrees of skill, the implications of piece-rates, of habitual overtime working, and of a standard wage. His treatment of unions was even-handed; the violence and intimidation associated with them in the public mind he deemed isolated and atypical. In general, he argued, combination developed a responsible approach and respect for democracy in working men, and the self-help aspects of unionism, the aid given to sick and unemployed colleagues, was commendable. This was consistent with views formed years earlier when a pupil at Fairbairn's in Manchester. 'I have mixed much with workmen. I worked with them side by side at the bench for nearly four years and during that time the great Engineers' strike occurred'.[106] On that occasion 'the masters stultified their cause by obstinate impolicy, and the men disgraced their order by acts of outrage'.[107] Yet on the whole he admired the workforce, and was respected by them in his turn although he had never accepted the prevailing wisdom that pupils court popularity by drinking with and otherwise emulating the workmen. Never elitist and always keen to learn, he saw the chance to be close to workers as a positive advantage of his profession.

> For the skilled artisan he had a great esteem, liking his company, his
> virtues and his taste in some of the arts. But he knew the classes too well
> to regard them, like a platform speaker, in a lump.[108]

Jenkin's conclusion was typically pragmatic: that as unions could and did improve the lot of their members, and as their abolition would be 'impolitic, undeserved and impossible', they should instead be regulated and operate with transparency in order to prevent any intimidation or misuse of power. He considered his own views to be 'heterodox', at odds with most commentators on the subject, but was anxious that Douglas did not reject his article on that account. He emphasized that the piece had been read carefully 'by some economical friends of mine, as well as by a secretary or two of some trade unions', that he himself was well read in economics and had had the advantage of long discussions with Alfred Austin.

---

[106] National Library of Scotland, Douglas papers, letter 112, Jenkin to Douglas, 10 February 1868.
[107] RLS, pp. xlix–l.
[108] RLS, p. xlix.

I tell you these things that you may be encouraged not to reject my writing because I differ on almost every point from the articles which have appeared in the *Edinburgh* and *Quarterly* [*Reviews*] and indeed from most writers in the daily and weekly press. I am afraid the legislature will make some horrible blunder in sheer ignorance of what artizans really want and of the rights of which they will really be found tenacious. English workmen are nearly inarticulate; they cannot explain themselves and they will not explain themselves. I have been quite excited by reading the questions and answers recorded by the Royal Commission from seeing how utterly unintelligible the question was to the workman and the answer to the M.P. They each think one another perfect brutes and if they could only have understood one another, each would have agreed with the other.[109]

Douglas complied with Jenkin's wish that the article be published quickly, in the hope that it might yet influence the Royal Commission. It appeared in March 1868, less than a month after Jenkin had finished writing it.[110]

The long article on trade unions is characteristic Jenkin, at once logical though rooted in realism, undogmatic yet eager for an answer. Commercial circumstances could not be dictated by laws of nature, but the engineer saw economic theory as a way of measuring and understanding: 'The ... view that somehow wages or prices are fixed by a law, is something like the idea that the strength of a beam is fixed by an equation.'

As this was published, Jenkin was working on his next and most celebrated contribution to economics, 'The Graphic Representation of the Laws of Supply and Demand, and their Application to Labour'.[111] This contained the first diagrammatic presentation of supply and demand curves to appear in English economic literature. Jenkin's graphs differed from subsequent practice in economics only in showing price on the horizontal axis and quantity on the vertical.[112] His next paper, 'On the Principles which Regulate the Incidence of Taxes', developed the graphic method to express effects of trade and property taxation.[113] Jenkin's final works on economics, 'The Time-Labour System: or

[109] National Library of Scotland, Douglas papers, letter 112, Jenkin to Douglas, 10 February 1868.
[110] Jenkin submitted the first half of the article on 10 February 1868, and it was accepted by return of post. It was published before 5 March: National Library of Scotland, Douglas papers, letter 112, Jenkin to Douglas, 10 February 1868.
[111] In Sir Alexander Grant (ed.), *Recess Studies* (Edinburgh 1870). Three letters from Jenkin to the economist W. Stanley Jevons in Manchester, dated March 1868 and addressed from the Duke Street office, show Jenkin's interest in Jevons's theory of exchanges: John Rylands University Library, Manchester, JA6/2/220–222.
[112] Whittaker 1947, p. 451.
[113] *Proceedings of the Royal Society of Edinburgh*, VII (1871–2), pp. 618–31.

how to avoid the evils caused by strikes' (1879–81) and 'Is One Man's Gain Another Man's Loss?' (1884) were unpublished during his lifetime. 'Time-Labour' defined differences between goods markets and labour markets and suggested improving the operation of labour markets through a system of guaranteed annual wages. 'Is One Man's Gain ...' used a form of closed circuit diagram to illustrate the exchange process and its results.[114]

How important was Jenkin's work in economics? His main contributions to the subject coincided with a seismic shift as the classical consensus among British political economists was replaced by a neo-classical school.[115] Stanley Jevons was most closely associated with the 'Marginal Revolution', and his *Theory of Political Economy*, published in 1871, provided the clearest and most comprehensive analysis at that time. Several other scholars, some belonging to the small British group of professional economists, others in France, Germany and Austria, and some altogether outside the ranks of mainstream economics, concurrently developed similar ideas and techniques.[116] Jenkin has been recognised as one of four major thinkers of the new movement, alongside Cliffe Leslie, Walter Bagehot and Jevons – a considerable feat for a non-specialist.[117] His industrial experience qualified him to relate theory through practical example, and mathematics provided a tool, a language in which to express economic concepts. The idea that he had read little other than Mill is mistaken.[118] He knew Malthus's ideas, and drew heavily on W.T. Thornton for the 'Supply and Demand' paper. British economists are said to have largely ignored continental ideas,[119] but Jenkin certainly had the opportunity to encounter leading European economic thinkers during the 1860s. In France there was a considerable overlap between civil engineering and the development of economic theory. Jules Dupuit, whose work in economics during the 1840s anticipated that of the British Marginalists a generation later, also wrote on water supply and was practising as a waterworks engineer during the time Jenkin worked on the Rouen waterworks project.[120]

There was also a brief dialogue between Jenkin and Stanley Jevons himself during 1868. Jevons had presented a paper entitled 'Brief Account of a General Mathematical Theory of Political Economy' to the British Association for the Advancement of Science in 1862,[121] which was published in the same volume as some of Jenkin's work on electricity, and was later expanded in the

---

[114] See Collison Black 1987.
[115] See Deane 1978, ch. 7; Hutchison 1973; Schumpeter 1954, ch. III.
[116] Deane 1978, p. 93.
[117] Hutchison 1973, p. 188.
[118] Brownlie and Lloyd Prichard 1963, p. 216.
[119] Hutchison 1973, p. 182.
[120] Thanks to Susan Morris for information about Dupuit and the suggestion of a connection with Jenkin's work in France.
[121] Deane 1978, p. 94.

Statistical Journal for 1867. Jenkin had not read this last piece before writing his 'Trade Unions' paper. When he sent Jevons a copy of 'Trade Unions' immediately after publication, asking for an opinion, Jevons responded with a copy of his own 1867 article. A short correspondence ensued about marginal utility – the idea that as further units of a commodity are consumed, their value to the individual diminishes.[122] This view, that value depended upon the feelings of the buyer rather than on any intrinsic quality of the good, was afterwards made explicit by Jevons in his *Theory of Political Economy*.

Jenkin's work has attracted praise for its lucidity and precision, and its understanding of the facts of industrial and commercial life.[123] As an economist, he has been described as 'highly distinguished'.[124] Yet the level of his influence is a different matter. For one thing it is difficult to establish just how novel Jenkin was, as a number of economists were separately pursuing similar lines of thought, and continental writers had already anticipated some of his ideas. The diagrammatic method, for instance, had been used, albeit in a much less developed form, by Cournot, Dupuit and von Mangoldt. While recent writers have given Jenkin credit for arriving at it entirely independently, there is really no clear evidence to establish how original his method was, other than a claim by the man himself.[125] The second problem in evaluating Jenkin's impact is that eminent contemporaries were less than generous in acknowledging him. It is not true to say that his contribution 'passed all but unnoticed'[126] but neither Jevons nor Alfred Marshall properly admitted it. Marshall gave both Dupuit and Jenkin 'but footnote recognition and this not in the right places'.[127] While Jevons admitted that his *Theory* had been hurried into print partly as a result of the 'Graphic Representation' article, he would not concede that Jenkin had offered anything original: 'From about the year 1863 I regularly employed intersecting curves to illustrate the determination of the market price in my lectures at Owens College...'[128] Jevons insisted that his mathematical theory of economics had been public since 1862, and that Jenkin's first treatment of the laws of supply and demand in mathematical language had not appeared until 1868, in the trades union article. Furthermore both had illustrated letters to each other in 1868 with curves, although Jenkin's 'Graphic Representation' paper in 1870 did not mention that Jevons also used graphs.[129] Jenkin himself was

---

[122] Jenkin's letters to Jevons are reprinted with explanatory notes in Collison Black 1977, p. 166 etc.

[123] Collison Black 1987, p. 1008.

[124] Hutchison 1973, p. 188.

[125] Brownlie and Lloyd Prichard 1963, p. 211, and also Schumpeter 1954, p. 1061, suggest that Jenkin was highly original in using this method.

[126] Schumpeter 1954, p. 838.

[127] Schumpeter 1954, p. 840.

[128] Brownlie and Lloyd Prichard 1963, p. 215.

[129] Collison Black 1977, p. 166.

convinced that he had started the 'system of considering political economical questions now in full use', and told Thomson that although Jevons claimed to have forestalled him, the British Association had printed 'only an unintelligible abstract' of the 1862 lecture, which did not amount to publication of the mathematical theory.[130]

Though undervalued at the time, Jenkin's work in economics did leave a lasting impression. Schumpeter has him as

> an economist of major importance, whose... papers form an obvious stepping stone between J.S.Mill and Marshall in four important respects: he was the first Englishman to discuss... demand functions; he both developed and applied to problems of taxation the concept of consumers' rent; he used diagrammatic representation, in principle, much as Marshall did later on; and he greatly improved the theory of wages, particularly in the matter of the influence of trade unions upon wage rates. In addition, like Sismondi but much more neatly, he suggested a time-labor system, essentially the 'guaranteed wage'.[131]

Jenkin was prominent among critics of the wages fund doctrine who forced Mill's recantation in 1869.[132] Mill conceded that this concept to which he had long adhered was artificial: 'The doctrine hitherto taught by all or most economists (including myself), which denied it to be possible that trade combinations can raise wages ... is deprived of its scientific foundation, and must be thrown aside'.[133]

Jenkin did not altogether abandon economics, as the later unpublished papers show, yet his most influential work appeared in just four years, between 1868 and 1872. There is no obvious explanation for his withdrawal. Jevons's *Theory of Political Economy* provided the subject with a new paradigm, and brought to an end the exciting and innovative period of change. Most likely Jenkin felt he had little new to contribute. Possibly he was vexed by Jevons's failure to acknowledge his work, and realized that as an outsider in the field he was not likely to be granted due credit. Yet Jenkin was not a conceited man, nor would he yield to vested interests if he felt that he had something important to add.

In the mid-1870s Jenkin turned his attention to public health, prompted partly by an exchange of correspondence in *The Times* during October 1876. He wrote to suggest various means of improving sewerage arrangements in

---

[130] Kelvin, Cambridge, J31, Jenkin to Thomson, 7 February 1883. Jenkin gives Jevons's name mistakenly as Symons.
[131] Schumpeter 1954, pp. 837–8.
[132] Hutchison 1973, p. 194.
[133] Quoted by S.G. Checkland, *The Rise of Industrial Society in England, 1815–1885* (London: Longmans, 1964), p. 415.

houses as a way of preventing disease.[134] This was not a completely new interest, for sewerage systems were a part of the degree course which he taught, and he had already contributed on the subject to the *Sanitary Record*. Yet it was a local tragedy which prompted more practical action. Friends of the Jenkins, a middle-class Edinburgh family, lost several children to an illness which was blamed upon the insanitary condition of their house.[135] Some of the best new houses in the city had potentially dangerous drainage systems. Jenkin's solution was a kind of co-operative, the establishment of a Sanitary Protection Association. The idea came to him from William Fairbairn's scheme in Lancashire, the Steam Users' Association, which aimed to reduce the number of boiler explosions through regular inspection. The Edinburgh Association cost one guinea a year to join, for which householders received a free annual check by a junior engineer, one of a number employed exclusively by the association. Anything above this inspection was to be paid for. Members would receive reliable advice as the association, unlike other contractors, had no reason to recommend unnecessary work or sell expensive patent devices. Stevenson explained it as 'a scheme of protection against the blundering of builders and the dishonesty of plumbers'.[136]

Jenkin acted as consulting engineer to the association without pay, rather, as he explained it, like a hospital for the poor where a leading physician would give his services free. Yet the Sanitary Protection Association was not for those in poverty. It was simple, pragmatic, popular – within a few months, there were five hundred subscribers in Edinburgh, and similar groups quickly formed in other British cities – but a guinea a year was beyond the reach of the poor. Jenkin's approach was to do what could be done, spreading the word about good practice in the hope that it would be widely adopted.[137] 'Fleeming believed we had only to make a virtue cheap and easy, and then all would practise it; that for an end unquestionably good, men would not grudge a little trouble and a little money, though they might stumble at laborious pains and generous sacrifices.'[138] Jenkin saw his association as complementary to the work of local councils, providing a service which was beyond their resources: 'My association in Edinburgh has done good work during the past year and all

---

[134] *The Times*, 19 October 1876.

[135] For this and further details of Jenkin's sanitary work, see Lieutenant Colonel Alexander Fergusson, 'Note on the Work of Fleeming Jenkin in Connection with Sanitary Reform', Appendix II in Colvin and Ewing (eds) 1887, I.

[136] RLS, p. cxxviii.

[137] Jenkin's lectures on the subject were published as *Healthy Houses* by David Douglas in 1878. His influence spread beyond Britain: he advised Alfred Ewing who was trying to persuade the Japanese authorities to adopt safer drainage systems: Ewing 1939, pp. 53–4.

[138] RLS, p. cxxix.

fear of any collision with local authorities is I think at an end. I think we supplement the local authorities in a very valuable way'.[139]

The 'Healthy Houses' scheme exhibited Jenkin's usual characteristics – energetically organized, imaginative yet uncomplicated, realistic in its objectives and far from dogmatic. Unlike Samuel Smiles, who had written on 'Healthy Homes' in the same year that the Edinburgh association was established, Jenkin resisted any urge to moralize by relating cleanliness to virtue.[140] His stance was moral rather than moralistic. The Sanitary Protection Association was a means of demonstrating the utility of scientific progress. With Colvin, Jenkin often discussed 'the advantages and disadvantages of science and mechanical discovery', taking a generally optimistic view of modern urban life, while accusing his friend of being 'a puling sentimentalist' in regretting the passing of a simpler rural existence.[141] The sanitary protection scheme would show that engineering could ease the burden on humanity by 'the application of science to human necessities'.[142] This idea was developed further in Jenkin's final paper on economics, 'Is One Man's Gain Another Man's Loss?', written in the year before his death. He concluded that when technology enables a few to supply the needs of a whole community, large numbers are set free to supply new wants. One man's gain does not of necessity mean loss to another.

As Jenkin's interest in political economy grew out of ideas on population during the 1860s, a parallel question simultaneously engaged him. By 1863 – before the move to Claygate – he had written a critique of Darwin's *On The Origin of Species*, first published in 1859.[143] Jenkin's essay did not appear until 1867, in the *North British Review*, and more than another year passed before Darwin discovered the author's identity.[144] It is certain that Darwin took Jenkin's arguments extremely seriously: 'Fleeming Jenkins [*sic*] has given me much trouble, but has been of more real use to me than any other essay or review', he wrote in 1869.[145] And shortly afterwards: 'Fleeming Jenkyn's [*sic*] arguments have convinced me'.[146] Jenkin himself told Thomsonthat his article had prompted revisions to Darwin's fifth edition, and that Darwin had acknowledged as much 'with his usual admirable candour'.[147]

---

[139] University of Edinburgh, Gen. 1729–32, Jenkin to J.B. Russell M.D., 20 January 1880.

[140] See Samuel Smiles, *Thrift* (London: John Murray, 1877), ch. XV.

[141] Colvin 1921, p. 158.

[142] Colvin 1921, p. 155.

[143] RLS, pp. lxvii–lxviii.

[144] Morris 1994, p. 313.

[145] Hull 1973, p. 302.

[146] Quoted by Morris 1994, p. 315.

[147] Kelvin, Cambridge, J31, Jenkin to Thomson, 7 February 1883.

Yet there is ambiguity in Darwin's allusions to Jenkin's criticisms, and the exact nature of Jenkin's contribution to evolutionary theory is open to argument. It was long assumed that it was on the mechanisms of evolution that Jenkin changed Darwin's mind. Specifically Jenkin argued that single variations could not survive being blended back into a general population which lacked their distinctive feature. Recent studies of the sequence of Darwin's revisions to his theory support the idea that he could not have read Jenkin's essay before modifying his views on single variations. Darwin's amended views appear in *Variations of Animals and Plants under Domestication*, published in 1868 but submitted for printing in November 1866, before the *North British Review* article appeared.[148] The influence of Jenkin, it seems, had exerted itself elsewhere in Darwin's thinking.

In fact it has been shown that before ever reading Jenkin's paper, Darwin had considered all the arguments which it contained.[149] Yet if Jenkin offered nothing new, why did Darwin praise him privately and so readily acknowledge his help in the fifth edition of *Origin of Species*? The answer lies in Jenkin's relationship with Thomson, Tait, and other physicists who were Darwin's most effective critics, and in his gift for bridging the gulf between areas of knowledge. One of Darwin's major problems was to establish geological time; he had to prove that the earth was sufficiently old to accommodate the slow process of evolution. Darwin's efforts to ascertain a framework within which evolution could be fitted, opened a debate which was to expose a chasm between methods and understanding in natural sciences, compared with those in mathematics and physics.

Thomson and Tait, separately and jointly, entered the argument on geological time. Thomson's papers which Jenkin had helped guide on to the British Association agenda in Manchester in 1861, on the cooling of the earth and the age of the sun's heat, were a part of that debate. Thomson thought that on the basis of the rate at which the earth cooled, there could not have been enough time available for evolution to occur. Jenkin and Thomson together had pondered these issues long before 1867: 'You will see a good deal of our old Arran talk in a N[orth] B[ritish] article on origin of species,' Jenkin wrote to Thomson at the time his critique of Darwin appeared.[150] The controversy, in which Thomson and Tait were ranged against T.H. Huxley and others, revealed bitter divisions between physicists who used mathematical methods, and geologists sceptical of mathematical evidence.[151] Darwin for a time apparently

---

[148] See Vorzimmer 1963, pp. 371–90; also Morris 1994, Hull 1973, pp. 344–50.

[149] See Vorzimmer 1963; also Hull 1973, p. 345.

[150] Kelvin, Glasgow, J65, Jenkin to Thomson, 23 May 1867. Thomson spent most summers in Arran and presumably had entertained the Jenkins there: Smith and Wise 1989, pp. 50–51.

[151] For a list of these publications, see the *North British Review*, L (1869), p. 406.

disregarded this dispute, which continued through the middle years of the 1860s, until Jenkin showed him that his entire thesis was threatened unless the issue was confronted.[152] Jenkin's role in this was not to produce original ideas, but to interpret and synthesize existing arguments, and to force Darwin to consider counter arguments potentially fatal to his theory. It seems that Jenkin did not have fundamental differences with Darwin, but in rehearsing arguments against evolutionary theory he reframed them in a way which enabled Darwin to turn them to advantage, making clearer some of the more ambiguous sections of *Origin of Species*.[153]

Thomson's recent biographers have given credit to Jenkin for moving along Thomson's own ideas, bringing 'the whole doctrine of energy to bear on the question of geological and cosmological time ... Apart from providing one of the most lucid explanations of the energy doctrines given at the time, Jenkin focused on the implications of Thomson's irreversible cosmos ...'[154] An idea was growing that the whole universe was evolving, following the Frenchman Sadi Carnot's concept of 'reversibility', and by implication its converse 'irreversibility', which had first been published in 1826. In general, processes are irreversible; from a final state an initial condition cannot be recovered for there are always changes that have a random quality, usually ending up as heat. Carnot's work was restricted to steam (heat) engines but the theory applied to almost anything – batteries, life, the stars, and information, for example. The connection between Jenkin's work and the irreversible cosmos showed a broadening interest in ideas of energy conservation, entropy and cosmical theories. His *North British* article linked Thomson's arguments, on the age of the sun and the cooling of the earth, and Darwin's theory, beyond anything that Thomson himself had attempted.[155] Significantly, when Peter Tait in 1869 summarized recent debate on geological time, Jenkin's critique of Darwin was identified as one of a small cluster of major articles.[156] Tait saw Jenkin's work in the context of that debate – not primarily as an article about single variations – and considered that Jenkin raised a fundamental geological question which geologists had failed to answer.

Ultimately geologists, despite their rejection of mathematical methods, proved to be more accurate than physicists in estimating the earth's age,[157] a result gratifying to supporters of Darwin. Thomson's assumption that the earth had been cooling since its formation was refuted by later discoveries,

[152] Morris 1994, p. 343.
[153] Morris 1994, p. 339.
[154] Smith and Wise 1989, pp. 536–37.
[155] Smith and Wise 1989, p. 580.
[156] 'Geological time', *North British Review*, L (1869), pp. 406–39. Tait is identified as the author in Houghton (ed.) 1966, p. 693.
[157] Hull 1973, p. 350.

particularly the finding that radioactive elements in the earth released enough heat to keep the earth's core molten. In a way the conclusion mattered less than the cross-disciplinary debate which Darwin had provoked. The argument was located where Jenkin was thoroughly at ease, on the cusp between fields of study. His talent was to interpret between specialists in different subject areas, and to translate technical arguments for the non-specialist. This was the man, after all, who had achieved a common scale of measurement for insulators and conductors. Reconciling the evidently incompatible was his speciality.

Jenkin had been born too late to be a polymath in the eighteenth-century sense. Branches of knowledge were becoming ever more clearly subdivided, and defined in terms of their membership. This sharper separation meant that the influence of an outsider, however sound or original their work, would never carry the weight of a specialist even in a relatively new discipline such as economics. But new definitions gave an opportunity for overview. As geologists and physicists discovered, there was a point at which subjects overlapped and compatibility needed to be established. Jenkin's prime offering to wider knowledge was a common language, the means of expression, communication, measurement.

It is tempting to dismiss Jenkin as lightweight because of his range of interests. Could this man be entirely serious? He spread himself too thinly to fulfil his potential in any field. And yet – whatever he did, he did well. He was widely admired, without having courted admiration. Our judgement of him filters through a modern culture of specialization which barely admits the value of breadth in thinking. With Jenkin, perhaps there is nothing to explain. It may be more eccentric, more questionable, to distil one's interests down to maths and golf, like Tait, and Jenkin loathed the narrowness of Tait. Ultimately it was a matter of character.

Stevenson provides a key to understanding Jenkin's wide interests, investing them with a religious significance:

> For Fleeming, one thing joined into another, the greater with the less. He cared not where it was he scratched the surface of the ultimate mystery – in the child's toy, in the great tragedy, in the laws of the tempest, or in the properties of energy or mass – certain that whatever he touched, it was a part of life – and however he touched it, there would flow for his happy constitution interest and delight.[158]

And while these subjects were weighty, gravitas and pomp were not Jenkin's style, for 'he was yet able, until the end of his life, to sport upon these shores of death and mystery with the gaiety and innocence of children'.[159] The desire to

---

[158] RLS, p. cxxxi.
[159] RLS, p. cxxxii.

understand meant that he chased tangents – which did not necessarily prove tangential.

As much a part of Jenkin's nature as breadth, was thoroughness. It was too convenient to accept a superficial explanation, for example in ascribing to inherent wickedness actions of which one disapproved: '...badness is such an easy, lazy explanation. Won't you be tempted to use it, instead of trying to understand people?'[160] Argument or discourse was the only means to progress, as Jenkin emphasized in his strong desire to be within a university, to be part of the group of professors. Brilliant conversation was more than a game, for to develop thoughts in one sphere was to advance in another. Although the polymath had disappeared, training in many of the new branches of knowledge remained *ad hoc* – in Jenkin's own world of electrical engineering every bit as much as in economics – yet an educated person was able still to contribute in other disciplines, in a way barely possible a generation later. Simultaneous discovery, where Jenkin's work kept pace with specialists in fields not his own, give proof of the level of his achievement.

Art and science, to Jenkin, were not mutually exclusive. He saw himself no more as a scientist dabbling in art than as an artist experimenting in science. He could not understand how Stevenson, with his antipathy for machinery, could consider himself a poet. To Jenkin, 'the struggle of the engineer against brute forces and with inert allies was nobly poetic'.[161] When he wrote to Annie on the joys of electrical experiment – 'What shall I compare them to? – A new song? a Greek play?' – it was without artifice.[162] True satisfaction came in the marriage of 'two devotions, art and science'.[163]

[160] RLS, p. cxxx.
[161] RLS, p. xlix.
[162] RLS, p. lxi.
[163] RLS, p. l.

Chapter 7

The Soul of the Enterprise[1]

Life in Edinburgh fell into an agreeable routine. At the centre of Fleeming Jenkin's domestic contentment was Anne, his 'heaven on earth'.[2] At first in Fettes Row, and from 1873 at 3 Great Stuart Street, Anne Jenkin, known to her three sons and all the family's friends as 'Madam', Figure 7.1, presided over a comfortable and busy house. The boys, taught at first by their mother, went on to prep school and later to the Edinburgh Academy. There were also dogs, 'domestic tyrants'[3], Jack the Irish terrier and Jenkin's own dog Plato who attended his lectures. The cat, Martin, became the subject of an article by Jenkin in *The Spectator*.[4] After a time Anne had her own lady's maid, with three housemaids and a cook making up the household.[5]

The grandparents too settled in Edinburgh. Alfred and Eliza Austin lived at Hailes. Charles and Henrietta Camilla Jenkin stayed for a time in Stirling – Never was there a place in which intellectual communication seems so impossible', complained Mrs Jenkin in 1872 – and had moved to the Edinburgh suburb of Merchiston by 1875.[6] There they were to spend the rest of their lives. Charles Jenkin, though still not in active service, had been promoted Captain in April 1861.[7] Henrietta Jenkin wrote occasional novels and reminiscences until her health began to fail soon after the move to Merchiston Place.

Jenkin resumed the country sports of his Borders childhood. He took on lease a hunting lodge at Glenmorven in the Highlands, extended it and staffed it with a gardener and a ghillie. There he shot, and fished for salmon, and began to learn reels and study Gaelic – one language to which he did not take easily.[8] He taught his boys these sports, and to swim and sail. In 1879 he bought a

---

[1]   Hempstead 1993, p. 204, quoting John Perry.
[2]   RLS, p. cliv, quoting Jenkin. See also Ewing 1933, p. 252.
[3]   Jenkin mss, Mrs Roscoe's recollections.
[4]   Ewing 1933, p. 259; RLS, p. cxvii; *The Spectator*, 20 August 1881. Martin's death was noted by Stevenson in May 1887: Booth and Mehew (eds) 1995, V, p. 413, Stevenson to Anne Jenkin.
[5]   Scottish Record Office, SC70/1/244, p. 817.
[6]   National Library of Scotland, Ms 4291, ff 177–180v., H.C. Jenkin to Mr Swayne, 19 January 1872; RLS, p. cxvi.
[7]   PRO, ADM 196/1, p. 164.
[8]   RLS, pp. cxv, cxx.

Figure 7.1: Anne Jenkin in Edinburgh, 1878 (Jenkin mss)

steam launch, the *Purgle*, to give his sons practical experience of engines. Frewen was the ship's engineer, Bernard the stoker, and Mrs Jenkin a crew member. In one notable episode the *Purgle* was taken down the west coast of Scotland through days of storms which endangered and discomfited them all. According to Stevenson, Jenkin, while relieved to have survived, was not deterred from sailing and saw the episode as positively beneficial for the family.[9] To take a risk was a healthy experience.

In other ways the Jenkins differed from many of their neighbours. Eminently respectable, moving in the most distinguished Edinburgh circles, there was yet a hint of eccentricity, or at least of a dissenting moral code.

---

9   RLS, pp. cxxi–cxxii.

Fleeming and Anne Jenkin had no time for snobbery and made a point of inviting to dinner a fellow professor and his wife who were shunned by polite society because of the wife's supposed low origins.[10] Alfred Ewing later described an 'atmosphere of distinction – intellectual, aesthetic, ethical' in the Jenkin house, the haven and oasis for the young Stevenson.[11]

The Great Stuart Street house was the scene of much of Jenkin's experimental work during the 1870s, with Alfred Ewing there first as assistant, later graduating to colleague and co-author. One joint paper of this period, 'On Friction between Surfaces moving at Low Speeds', published in the *Philosophical Transactions* of the Royal Society for 1877, is recognised as a particularly important contribution to science. It investigated the coefficient of friction between metals, and between metal and wood, at very low speeds in relationship to the velocity of the rubbing surfaces.[12] In 1877 at Thomson's behest Jenkin applied to the Royal Society, which administered government grants to scientific research, for support in further experiments on friction. He was unwilling to continue meeting all the costs personally, though he asked for only fifty pounds.[13] This small request was at first turned down, causing Thomson to write to Sir George Gabriel Stokes, the Royal Society's secretary:

I don't know if my being a member of the Committee could have made a difference in the result, but I think I could certainly have shown good reason for granting the sum applied for. Jenkin, with the assistance of J.A. Ewing, has made experiments on friction at very low speeds, by a beautifully devised dynamical method, which they have carried out *very carefully* in a first investigation which has already brought out remarkable results.[14]

There was evidently a change of heart and Jenkin was offered the money. In the event Ewing's move to Japan meant a delay, so that the grant was not claimed that year. Jenkin continued with a new assistant:

The experiments are begun, and I can already say that this will yield trustworthy results ... The scope of the experiments is the complete connection between friction of rest and friction between two bodies in relative motion, taking into account the time during which given surfaces have been in contact. I hope in this way to bring all we know about

---

[10] Jenkin mss, Mrs Roscoe's recollections.
[11] Ewing 1933, p. 250.
[12] Obituary, *Institution of Civil Engineers Proceedings*, p. 375.
[13] Royal Society, *Government Grant Applications, 1877–83*, 1, pp. 36, 52. Paul Byrne of the Royal Society supplied information on government grants.
[14] Wilson (ed.) 1990, II, pp. 429–30, Kelvin to Stokes, 21 March 1877.

friction into a consistent whole, and to know that changes which have hitherto been considered abrupt are really continuous.[15]

Jenkin's most remarkable scientific paper, it is generally agreed, was 'On the Application of Graphic Methods to the Determination of the Efficiency of Machinery', published in two parts in the *Transactions of the Royal Society of Edinburgh* for 1876–78.[16] This laid the foundations of a new dynamical theory of machines, and was rewarded with the Society's highest award, the Keith Gold Medal, in 1879. Jenkin's hypothesis grew out of Reuleaux's 'Kinematics of Mechanism', which is based on a premise that every mechanism is a self-contained whole, a complete chain of connected parts. Jenkin cut out superfluous detail in Reuleaux to develop an analysis incorporating dynamics, so that Reuleaux's assemblage of parts became a powered, working machine. Jenkin acknowledged that his work built upon that of Reuleaux, but his paper was thought so original that without the author's indication, Reuleaux's influence would not have been suspected.[17] Jenkin's model produced the means of understanding a complex problem, that of calculating precisely the waste by friction which is lost as power passes through a machine. Jenkin himself was pleased with his achievement: 'I took the medal with a good conscience, for I confess to being proud of that paper', he told Ewing.[18]

Less immediately gratifying were Jenkin's efforts with heat engines, lasting for a decade from the early 1870s. He saw a potential economy if engines could burn their fuel internally and achieve higher temperatures. He experimented with a regenerator on gas engines, and with engines burning coal in a closed chamber under compression. In 1874 a patent was taken out for an engine upon which 'fruitless and disheartening experiments' continued for years afterwards.[19] Eventually, in 1881, there were three further patents, for engines of the Stirling type but which burned fuel within the displacement chamber.[20] Once prototypes had been built Jenkin had to concede that he was working on too small a scale to produce operational engines.[21] To his credit, Jenkin had persevered with the problem despite the endless difficulties. Ewing helped him with it until moving to Japan in 1878, after which it became something of a standing joke. 'You will be amused to hear that with Jamieson's assistance I still potter on at the old heat engine ... ' Jenkin wrote to Ewing in

---

[15] Royal Society, *Government Grant Applications, 1877–83*, 1, p. 249.
[16] In particular Sir William Thomson, in the obituary in the *Royal Society Proceedings*, singled out this work as Jenkin's most significant publication.
[17] Obituary, *Institution of Civil Engineers Proceedings*, pp. 375–76.
[18] Ewing 1933, p. 273.
[19] Obituary, *Institution of Civil Engineers Proceedings*, pp. 370–71. The patent is 2441 for 1874.
[20] Patents 1078, 1130 and 1160 for 1881 were taken jointly with A.C. Jamieson.
[21] Obituary, *Institution of Civil Engineers Proceedings*, pp. 370–71.

1879.[22] His new assistant was Andrew Jamieson (b.1849), an *alumnus* of Aberdeen University who also worked for Thomson and Jenkin on telegraph projects, and later became principal of the Glasgow College of Science and Art.[23] Three patents were taken out for 'Caloric Motor Engines', jointly with Jamieson, in 1881. Jenkin was still occupied on the project in June 1882, when Ewing wrote to him:

> So, Jenkin, you are working on the heat engine again! I hope you may get the better of it at last, if only to gratify my spite. I have nothing but unpleasant memories of that engine; it never inspired me with a spark of hope; I used to trudge down the Canongate daily with a perfect confidence of failure, and I was never disappointed.[24]

Jenkin did not entirely give up on the heat engine – there was another patent in 1884 – but gradually it receded into the background. The nineteen United Kingdom patents which Jenkin took out in the final three years of his life related mainly to other interests – electrical transportation, gearing and power transmission.

By 1870 the partnership with Thomson and Varley, while continuing to pay handsomely, required less personal involvement. When Jenkin took the Edinburgh chair he and Thomson set up another partnership, to act as consultant engineers to undersea telegraph companies. They supervised cable-making and laying on the Western and Brazilian, the Platino Brasileira, the West India and Panama, and the Mackay-Bennett Atlantic lines, and other lesser projects.[25] Much of the work could be delegated to assistants. Some were students such as Ewing, who from the age of seventeen had spent his summer vacations representing the partners in London cable factories and later on expeditions in Brazil.[26] Other employees were taken from different backgrounds, such as Sidney F. Walker, a gifted mathematician who had been a naval cadet before attending telegraph school, appointed by Thomson and Jenkin to work for them at Hooper's Telegraph Works at Millwall.[27] William Ayrton, then of the Indian Government Telegraph Service, a former student of Thomson, represented the partners at the London factory where the south American cable was being manufactured over the winter of 1872–73.[28]

---

[22] Ewing 1933, p. 273.
[23] *Electricians' Directory* (1885), p. 108.
[24] Ewing 1933, p. 256; Ewing 1939, p. 65.
[25] Obituary, *Institution of Civil Engineers Proceedings*, p. 367.
[26] Bates 1946, p. 3. See also Chapter 4, above.
[27] *Electricians' Directory* (1885), p. 133.
[28] *Electricians' Directory* (1885), p. 88.

In July 1872 Jenkin was in Carcavelos (Lisbon), apparently negotiating terms for a telegraph to Shanghai.[29] By that time Thomson and Jenkin had been appointed engineers to the Great Western Telegraph Company, established by a group of investors connected with Hooper's Telegraph Works, with the intention of laying a cable from Land's End to Bermuda, and lines from there to New York and *via* the Virgin Islands to Brazil.[30] This project underwent radical change early in 1873, with the Great Western company wound up and replaced by a plan to take the line from Portugal to Pernambuco, now called Recife, in Brazil. The Telegraph Construction and Maintenance Company, which held Brazilian concessions and was keen to avoid further competition on the Atlantic routes, had been instrumental in the new scheme.[31] The project was taken over by a new company, the Western and Brazilian Submarine Telegraph Company, with Pender, Gooch, and Anderson among its board members.

Despite Gooch's previous hostility, Jenkin and Thomson were retained as engineers. The partners embarked for Lisbon, *en route* to Brazil by way of Madeira and the Cape Verde Islands, in June 1873.[32] They were on board a new cable ship, the design of which they had jointly supervised during its construction on Tyneside during the previous winter. The *Hooper*, second only to the *Great Eastern* in size and the second ship built specifically to lay cables, was fitted with hydraulic side thrusters at the stern for maximum manoeuvrability.[33] Pernambuco was reached on the last day of July, after 2500 miles of cable had been laid, although a fault had developed and it was another year before the line finally opened.[34] Cable-laying was to continue down the south American coast but Jenkin did not return to Brazil. Instead Ewing took charge of the extensions to Rio de Janeiro and on to Montevideo in Uruguay over the following two seasons, with Jenkin confidently rebuffing the protests of Hooper that a nineteen year-old was too young to be responsible for such a project.[35] A stark reminder that cable-laying still carried great dangers came in November 1874 when the *La Plata*, on its way to Brazil, went down in the Bay of Biscay with the loss of sixty lives.[36] Ewing was not on board but a number of Jenkin's associates perished, including Frederick Henry Ricketts, who had

[29]   Kelvin, Cambridge, F9, Jenkin to Thomson, 18 July 1872.
[30]   Wilson (ed.) 1990, pp. 372–74, Thomson to Stokes, 14 October 1872; Barty-King 1979, p. 48.
[31]   Haigh 1968, p. 136.
[32]   RLS, pp. cxi–cxiii.
[33]   Smith and Wise 1989, p. 744; Haigh 1968, p. 78.
[34]   Barty-King 1979, p. 49.
[35]   Ewing 1939, p. 30.
[36]   Ewing 1939, p. 31.

taken out a joint patent with Jenkin in 1873 on V-pulleys for power transmission.[37]

The year 1874, when Ewing assumed practical cable-laying responsibilities, also marked a watershed in the Thomson-Varley-Jenkin patent marketing partnership. In 1874 a licensing agreement with the Eastern Telegraph Company put the siphon recorder and the partners' other instruments into routine use around the world, guaranteeing a further generous annual payment to the three.[38] Thomson told Jenkin in 1881 that their main patents were bringing in £3000 annually from the Eastern Telegraph Company, £2100 from the Eastern Extension and £1500 from the Anglo – this in addition to the several thousands a year which the Thomson-Jenkin consulting partnership was earning.[39] While there were new patents and occasional controversies with instrument licensees after 1874, most business was routine.[40]

With the technical and commercial problems of submarine telegraphy at last settled to his satisfaction, Jenkin started to investigate new electrical applications, particularly relating to communication. After telegraphy the next challenge for electrical engineers was to develop a practical system of lighting. Although hardly a novel idea, electric lighting was not feasible until economic methods of generating and distributing power had been developed. Among those actively working on solutions were Thomson, who gave evidence to the House of Commons Select Committee on Electric Light in 1879, and Siemens.[41] Jenkin, although interested, had a peripheral role: discussing the subject with Thomson, acting as consultant to the council in Great Yarmouth on electric lighting, writing a short article for *The Electrician*, producing a circular for lighting exhibitors in London in 1884.[42] By observing how lighting systems developed during the 1870s, and particularly how the weight of public expectation affected events, he learned the need for caution during his later telpherage trials: 'Our hopes of rapid progress were raised too high in the matter of lighting, and the check given to enterprise in one application of

---

[37] Ricketts, born in 1837, had been a pupil of Liddell and Gordon and was later employed by Forde and Jenkin as superintendent of the Rouen waterworks scheme, and helped them build scientific apparatus. At the time of his death he was working for Siemens. See *Proceedings of the Institution of Mechanical Engineers* for 1875, p. 30. Thanks to Keith Moore of the IME archives for this information.

[38] Sloan 1996, p. 195.

[39] Smith and Wise 1989, p. 705.

[40] There were sometimes problems with licensees: see for instance Kelvin, Cambridge, V10, S87 and LB4.

[41] Smith and Wise 1989, pp. 712–22.

[42] A copy of this leaflet is in the New York Public Library, Wheeler collection. Kelvin, Cambridge, J30, Jenkin to Thomson, 8 January 1883, refers to the Great Yarmouth commission.

electricity has acted as a deterrent to experiment in the new field.' Impatience, he concluded, must be reined in.[43]

In 1877 and 1878 two electrical novelties from the United States seized Ewing's attention and took his and Jenkin's investigations in a new direction. Jenkin's fascination with the telephone, and even more with the phonograph, related not to their mechanical or electrical construction, but was bound up with their potential as instruments for further scientific research. Rather than applying science to advance technology, Jenkin's idea was to use new technology to extend an understanding of natural phenomena. There is also a clear connection between his enquiries into vowel sounds – the roots of articulate language – and his interest in drama and articulation.

The telephone, the first of these instruments to burst upon public consciousness, was a source of fascination for Jenkin's submarine telegraphy colleagues.[44] In 1877 Alfred Rosling Bennett (1850–1928), formerly a telegrapher in India and the Middle East, made his name as a pioneer of telephony when he 'carried out the exhibition of Mr Cromwell F. Varley's musical telephone at the Queen's Theatre, Long-acre, which was joined for the purpose to the Canterbury Music Hall by two overhouse wires – the first telephone wires ever erected in this country'.[45] Varley had been working on a device which he called the cymaphen, a means of transmitting musical notes by wire.[46] The exiled Scot Graham Bell sent four telephones from America to Sir William Thomson, which Ewing rushed to Glasgow to see. Thomson allowed him to borrow them, and thus it was that the first public use of the telephone in Scotland took place at a church bazaar in Dundee, where the new devices attracted hundreds of visitors at half a crown a ticket. Afterwards Ewing linked the receivers to a length of telegraph wire and was able to transmit music which could be heard ten miles away. He predicted a day when there would be private wires linked to central offices through which housewives could order shopping, 'with all that fullness of verbal explanation which ladies are supposed to acquire'.[47]

Robert Louis Stevenson's cousins David and Charles, graduates of Jenkin's course who were to take over the family business as engineers to the Northern Lighthouse Board, had seen the first public demonstration of Bell's telephone in Philadelphia in 1877. Alexander Graham Bell (1847–1922) was

---

[43]   'Telpherage', *Good Words* (1885), p. 139.
[44]   For example W.H. Preece, who from 1877 was Electrician to the Post Office, published a number of papers on telephony: *Electricians' Directory* (1885), pp. 120–21.
[45]   *Electricians' Directory* (1885), pp. 89–90. Bennett's career is described in an obituary notice in the *Journal of the Institution of Electrical Engineers*, 66 (1928), pp. 1233–34.
[46]   See *Journal of the Institution of Electrical Engineers*, 71 (1932), p. 961. Tim Procter provided information about Bennett and Varley.
[47]   Ewing 1939, pp. 43–44.

also an *alumnus* of Edinburgh University, although he had left in about 1866, before Jenkin's arrival. At University College, London, between 1867 and 1870 Bell studied anatomy and physiology rather than Jenkin's engineering course, and there is no evidence that Bell and Jenkin met.[48] Charles, the more inventive of the Stevenson brothers, was taken by another device, known as Edison's electrical chemical telephone, on display in London in 1879.[49] Thinking particularly of remote lighthouses, he began to consider the possibility of a system which worked without wires or cumbersome equipment. Hertz had already proved that electrical waves could pass through anything, but Charles Stevenson's notion of wireless communication was novel. He experimented on it constantly from 1880, sometimes working along with Post Office engineers, and it appears that he achieved a breakthrough in 1892 when radio signals were transmitted and received over a distance of two miles in Edinburgh. The Lighthouse Commissioners refused the scheme a trial on grounds of cost, and Stevenson's work was eclipsed by that of Marconi within two years.[50]

Jenkin and Ewing were even more enthusiastic about Edison's phonograph, news of which reached them in 1878. It occurred to Jenkin that this new device could do for Edinburgh University cricket club finances what the telephone had for the Dundee church. As it was not possible to obtain a phonograph in Britain, he and Ewing built two from a description given in *The Times*.[51] The machines recorded and replayed speech or music. Ewing took one to show to his students at the Watt Institute:

> He placed on the demonstrating table something that resembled an ordinary morse telegraph instrument; a tube into which the operator spoke; a small, grooved barrel wrapped round with tinfoil on which were received the indentations; and a hand-wheel which turned the cylinder slowly, whilst the voice recorded into the mouthpiece. As the timing was rather irregular, so the reproduction of song was a travesty, and that of speech seemed to emphasize all the least agreeable peculiarities of the speaker's voice. To hear oneself as others heard one, in the phonograph's interpretation, was a shock to the most complacent vanity.[52]

Queues of people paid for this experience at the cricket ground bazaar, and to attend lectures on the phonograph by the professor. Apart from a few visitors who 'were deaf, and hugged the belief that they were the victims of a

---

[48] See the *Dictionary of National Biography*, 1922–30, for Bell.
[49] Mair 1978, pp. 219–20. The Stevenson brothers David (1854–1938) and Charles (1855–1950) graduated in 1875 and 1877 respectively.
[50] Mair 1978, pp. 230–31.
[51] RLS, pp. cxxx–cxxxi.
[52] Ewing 1939, p. 44.

new kind of fancy-fair swindle', the device was a triumph.[53] But apart from its use as a money-raising novelty, and a source of amusement to Stevenson and William Hole – 'with unscientific laughter, commemorating various shades of Scottish accent, or proposing to teach the poor dumb animal to swear'[54] – Jenkin saw other potential in the phonograph. Ewing had noted that through it 'the consonants were weakened, the vowels strengthened'.[55] The 'immensely entertaining toy' enabled Jenkin and Ewing to carry out research into the nature of spoken sounds.[56] One machine had been raffled, the other was kept by Jenkin to record dozens of different voices. A joint paper, 'The Harmonic Analysis of Vowel Sounds' was read before the Royal Society of Edinburgh, and six articles on the subject by Jenkin and Ewing appeared in 1878, including one in *Nature*.

In 1879 Jenkin applied again to the Royal Society for a grant of fifty pounds, this time to construct 'apparatus for investigation of vowel sounds by means of the phonograph'. He explained that his investigations to date had examined the 'harmonic constituents' of the vowels *o* and *u*, and thrown some light upon *a*. In order to complete the research on *a*, and look at the sharper sounds *e* and *i*, a different method of working was needed.

> In the new method I propose to reverse the former process and to investigate the sound produced by cams cut so as to give different harmonics combined in very numerous ways. The idea is an old one, but I think that the apparatus I have designed will enable the cams to be cut with great accuracy, very cheaply, and in great variety; also that I have designed machinery by which the cam will give motion to a vibrating or speaking disc with perfect definiteness.[57]

It does not appear that the Royal Society supported Jenkin. After 1878 there were no further publications resulting directly from this research and it seems that the second stage of the research did not take place.

In his papers on Mrs Siddons, which were written at about the time of his interest in the phonograph, Jenkin missed the apparently obvious point that the machine could be a means of recording the work of great actors. This omission may have been a reflection of the very poor quality of recordings on their home-made phonograph – Ewing shouting into the mouthpiece at full volume reproduced only as 'the feeblest pianissimo – the merest echo'.[58] Even in 1915 the quality of reproduction was not outstanding, for when Brander Matthews

---

[53]   RLS, pp. cxxx–cxxxi.
[54]   RLS, p. cxxxi.
[55]   Ewing 1939, p. 44.
[56]   Ewing 1933, p. 269.
[57]   Royal Society, *Government Grant Applications, 1877–83*, I, p. 163.
[58]   Ewing 1939, p. 45.

reviewed Jenkin's work on Siddons he could only speculate that in future, 'the phonograph may preserve for us the voice of an honoured performer' while adding that 'at best, these will be but specimen bricks, and we shall lack the larger outlines of the performance as a whole'.[59] But there is no doubt that Jenkin's fascination with the phonograph connected to his continuing interest in drama and speech. In 1882 he proposed to the Watt Institution directors that they should offer a class in elocution, and he found a teacher who could give lessons in 'vocal physiology and the Art of Reading and Speaking'.[60] The premature ending of Jenkin's research into vowel sounds stemmed not from a loss of interest, rather from practical difficulties in continuing. Ewing, who had given impetus to the scheme, moved to Japan in 1878, and Jenkin was over-stretched with other responsibilities.

In 1882 came the last great project, a system of electrical transportation which he called telpherage − 'the telpherage schemes and dreams (if dreams they were) which so absorbed his latter years'.[61] Jenkin had not lost his ability to enthuse, to be captivated by new possibilities, to attack a problem with energy and originality. After all, he was not yet fifty. 'Far on in middle age', said Stevenson, 'when men begin to lie down with the bestial goddesses, Comfort and Respectability, the strings of his nature still sounded as high a note as a young man's.'[62] For some time he had been thinking about electrically-powered transport, 'doing on a large scale with electricity what had previously been done on a small scale with pneumatic tubes'.[63] The stimulus to turn his abstract ideas into something workable came from a lecture on electric railways given by William Ayrton and John Perry to the Royal Institution in 1882. They had worked out a system for applying an absolute block to prevent collisions, and demonstrated it with a working model. Using this discovery it seemed possible that electrical transport could run automatically, without the need for drivers, guards or signalmen.[64] Jenkin approached Ayrton and Perry, both then professors at the City and Guilds of London Technical College in Finsbury, with a proposal to go into partnership. He recognised that there were many possible ways to develop electrical passenger and goods transport, and that electrical transport would not usurp steam power for most applications. Electric railways were already operating; trams were run by battery. Ayrton and Perry had also proposed a system of electric haulage on canals. Jenkin, writing for a general readership in *Good Words*, realised that electric transport was seen as utopian, 'but the rate of progress during the last fifty years is

59   Matthews (ed.) 1958, p. 74.
60   Heriot-Watt University Archive, SA1/2 Minute Book of Directors, 14 March 1882.
61   Yale, Beinecke, 4390, Colvin to Stevenson, 13 September 1887.
62   RLS, p. cxxxvii.
63   Obituary in *Nature*, XXXII, 18 June 1885, p. 153.
64   Obituary in *Nature*, XXXII, 18 June 1885, p. 153.

encouraging for the future'. His own scheme had advantages in certain situations.

> The aerial Telpher line described seems suitable to half-settled countries, where roads, railways and canals are not as yet constructed, and where the traffic is not sufficient to warrant their construction. It will climb steep hills and go round sharp curves., and is therefore suited for broken, hilly ground. It may also find an application wherever it is important not to interfere with agriculture, and where it is necessary to remove the conductors from possible interference by men or beasts.[65]

Figure 7.2: Fleeming Jenkin in his laboratory in 1884, from an engraving by William Hole published in *Quasi Cursores*

He saw further possibilities – of running lines out to sea to load and unload ships, of using the telpher to feed into conventional railways, and even driving machinery on remote farms by attaching flexible conductors to the telpherage cables, Figure 7.2.

---

[65]   'Telpherage', *Good Words* (1885), pp. 138–39.

Telpherage, the name devised by Jenkin, is a slightly inaccurate rendering of the Greek *to carry far*. The telpher line was a flexible conductor of electricity, supported on posts, which was also strong enough to carry an electric motor and a train of light trucks or buckets hanging below.[66] The line was in short sections, electrically distinct, and the train, longer than a single section and itself a continuous conductor, bridged the interval between sections as it passed between them. A fixed engine and dynamo could be placed at any station, and the slow-running trains did not need attendants on board as speed, shunting and braking were controlled automatically.[67] Jenkin had developed and simplified Ayrton and Perry's block, so that any train travelling too close behind another would reach a section of line temporarily deprived of current, and would stop automatically.[68] The automatic working of telpherage, along with its safety features, Jenkin saw as selling points. He knew, though, that it would be successful only if it were an economical alternative to other means of transport. 'These lines must be cheap or their use will be restricted to certain exceptional cases'.[69]

The inspiration for telpherage had come to Jenkin from a popular myth that a pair of boots could be sent to a friend by hanging them on a telegraph wire.[70] It was in essence the telegraphing of goods. There was another similarity with telegraphy – telpherage was a complex mechanical and electrical system which required painstaking innovation in every detail. Rather than take up a partially developed project, as Jenkin had with deep-sea telegraphy, this was an opportunity to devise an entire scheme from scratch, and Jenkin, though advised and helped by partners and assistants, took full authority. 'His inventive power is described by his assistants as wonderfully active and prolific, and he had energetic characteristics which only seldom accompany inventive genius ...'[71] Every aspect of telpherage threw up problems to which Jenkin worked out solutions. 'His originality and powers of suggesting devices for overcoming difficulties were prolific, as is shown by his manifold patents in connections with the details of telpher lines'.[72] Three patents were taken out in 1882, and more in the following years.[73] The project took up more and more of Jenkin's energy. He wrote to Ewing in the summer of 1882: 'I am collaborating with Ayrton and Perry in a big locomotion scheme whereof more soon. Gas-engine

---

[66] For technical details of the system, see Hempstead 1993; also Hempstead 1994.

[67] Obituary, *Institution of Civil Engineers Proceedings*, p. 372.

[68] 'Telpherage', *Good Words* (1885), p. 138. This article provides a detailed explanation of the system as it operated late in 1884.

[69] 'Telpherage', *Good Words* (1885), p. 139.

[70] Hempstead 1993, p. 197, quoting Jenkin's address to the Glasgow Science Lecture Association, 1884.

[71] Obituary, *Nature*, 18 June 1885, p. 153.

[72] Obituary, *Journal of the Society of Telegraph Engineers*, p. 348.

[73] Hempstead 1993, p. 200.

drags along slowly: its nose is put out of joint by this new electrical affair.'[74] By December he was 'desperately busy over Telpherage'.[75] He was assisted by Archibald Campbell Elliot, who was to receive Edinburgh's first engineering doctorate in 1888.[76] Gordon Wigan, a barrister, son of a well-known actor, who shared Jenkin's interest in drama and who became a founder director of the Telpherage Company, was also helping Jenkin at this time, perhaps with patent applications.[77] Working with the telpherage partners was not proving easy for Jenkin, and he planned to introduce other trusted engineering associates when the time was right. He wrote in January 1883 to Thomson:

> I wish I could get time to tell you about Telpherage, for I want to bring you in as Consulting Engineer, doing some work gratis at first with a view to subsequent position as Director or Consulting Engineer if experiments are successful – I expect Clark and Forde to be Engineers on the same terms ... I could arrange all this nicely if it were in my own hands but Ayrton and Perry are prickly customers.[78]

A letter addressed from the Savile Club to Stevenson in the summer of 1883 hints at overwork and exhaustion. Jenkin had been unable to join his wife and sons for their annual holiday: 'I am working in London *en permanence* – telpherage and nest gearing – all the family are at Glenmorven. They say they want me – I want them and the peace.'[79] The nest gearing project was intended to improve the efficiency of machines where power was transmitted by rolling contact; rollers were grouped symmetrically in 'nests' to reduce friction. This idea Jenkin patented in 1883.[80]

Jenkin's presence in London was necessary as he was overseeing telpherage experiments in Hertfordshire. In March 1883 the Telpherage Company Ltd. had been incorporated to acquire Jenkin, Ayrton and Perry's telpherage patents, ultimately numbering more than a dozen.[81] The three engineers were paid with about 200 'B' shares, valued at ten pounds each. Additionally they had individual investments in 'A' shares. Jenkin held as many as 100 of these, and also allocated company stock to friends, sometimes without their knowing.[82] The total capital was £50,000, and the shareholders' list for 1884 includes William Thomson as well as Clark, Forde and Taylor,

[74] Ewing 1933, p. 256.
[75] Ewing 1933, p. 272.
[76] Jenkin and Elliot took out a joint patent for telpherage in 1884.
[77] Matthews (ed.) 1958, pp. 72–73.
[78] Kelvin, Cambridge, J30, Jenkin to Thomson, 8 January 1883. Both Ayrton and Perry were former students of Thomson.
[79] Yale, Beinecke, 4990, Jenkin to Stevenson, 3 August 1883.
[80] Obituary, *Institution of Civil Engineers Proceedings*, p. 371.
[81] PRO, BT 31/3135/18071.
[82] RLS, p. cliii.

and Charles Ernest Spagnoletti, shortly afterwards president of the Society of Telegraph Engineers.

The chairman of the Telpherage Company, and a key figure in furthering the necessarily expensive development work, was Marlborough Robert Pryor (1848–1920). The trials took place upon his estate at Weston Park, north of Stevenage. Pryor, who had been a fellow of Trinity College, Cambridge, until forced to relinquish the position when he married, was subsequently a merchant trading with South America and chairman of the Sun Life Assurance Company.[83] He lived at Weston Park, where he was known as a progressive farmer, from 1870 until his death in 1920. The telpher line there was used to carry agricultural produce across fields.[84] Jenkin took charge of the experiment, immersing himself in the work 'with even more than his usual inventive industry, executive ability and determination to overcome all obstacles'.[85] The trials lasted for more than a year, during which 'many forms of rope, rod, post, saddle, locomotive and truck, were erected and tried'. Jenkin summarized the result himself:

> Finally a short line, 700 feet long, was constructed which re-entered on itself, so that trains could be run round and round. The line consisted of ten 50-foot spans bridged by flexible round rods of steel five-eighths of an inch in diameter, joined by two semicircular ends each 100 feet long, made of stiff angle iron, supported on posts about 10 feet apart.[86]

Two trains, each weighing about a ton and travelling at four to five miles an hour, could run simultaneously. The system worked perfectly during a two-week public display in 1884. Further improvements to the braking system were tried on another short line which ran at Erith before being exported to South America.

By the end of 1884, the telpherage scheme had completed its transformation from abstraction *via* novelty plaything to potentially viable system. Jenkin knew that it needed a proper commercial trial in Britain in order to prove its worth.[87] In January 1885 he was able to add a triumphant footnote to his 'Telpherage' article in *Good Words* – a contract has been arranged by which the first British line will be erected at Glynde for the Sussex Portland Cement Company – to deliver 150 tons of clay per week'.

---

[83]   Venn (ed.) 1947, V, p. 212.
[84]   Information on Pryor was kindly provided by the Hertfordshire County Record Office. See E.V. Scott, 'The Pryors of Weston', *Hertfordshire Countryside*, 26, no.150, October 1971.
[85]   Obituary, *Institution of Civil Engineers Proceedings*, p. 372.
[86]   'Telpherage', *Good Words* (1885), p. 133.
[87]   'Telpherage', *Good Words* (1885), p. 139.

Rather than relieve the pressure on Jenkin, this commercial agreement added to his anxieties. The Cement Company had been offered generous terms in order to secure their participation in the trial. They would pay for the haulage, provided that 150 tons a week was transported from their Glynde clay pits to the local railway sidings with little delay. Otherwise the Telpherage Company, which was to pay the cost of the line, was subject to financial penalties. The Cement Company also retained a right to have the telpher dismantled after a set period. It was imperative to the future of the Telpherage Company, and to the system of telpherage itself, that the Glynde experiment succeeded.[88]

For the last six months of Jenkin's life he supervised the manufacture of electric locomotives and the building of the line, as well as solving innumerable mechanical problems.

> He persevered with endless ingenuity in carrying out the numerous and difficult mechanical arrangements essential to the project, up to the very last days of his work in life. He had completed almost every detail of the realization of the system ...[89]

The line at Glynde opened with much publicity in October 1885, four months after Jenkin's death. John Perry had succeeded him as engineer. Alfred Ewing, then professor at Dundee, had not been able to attend the opening of the line but reported to Thomson that Jenkin's sons had been present 'and were much pleased with it'.

> As to the Telpher line, I believe the opening was very successful. So far as I know, you are quite right in thinking Jenkin had actually worked the thing out to the minutest details before he died. Indeed, everything or nearly everything was made or ordered by that time. Some changes of plan have been made since − or rather change of detail. Jenkin's rods (forming the line) had peculiar forged ends; but it was found that they were liable to break at a particular section near the end, and new rods had to be designed. I believe that practically every detail of the locomotive, trucks and so on is Jenkin's. Perry, I know, had to complete the specification of a governor which Jenkin left unfinished. One change was the mode of emptying the 'skeps'. Jenkin's scheme had been to empty them (at the side of the railway) into hoppers which discharged into the railway trucks. Perry carried the telpher line over the railway, and emptied the skeps directly into the trucks beneath them.[90]

---

[88]   Hempstead 1993, n. 5.
[89]   W. Thomson, 'Obituary Notices of Fellows Deceased', *Proceedings of the Royal Society*, XXXIX (1885), p. ii.
[90]   Kelvin, Cambridge, E114, Ewing to Thomson, 27 October 1885.

The engineering work impressed some members of the press who were at the opening, although in general judgement on its commercial possibilities was reserved.[91]

The Jenkin family, eager to continue the telpherage scheme, wanted their most promising engineer, the middle son Frewen, to become involved. Frewen had gone up to Trinity College, Cambridge, in 1883 after studying engineering under his father at the University of Edinburgh.[92] Anne Jenkin told Thomson that Frewen was 'gradually, so as not to interfere with his work – making himself acquainted with the telpherage patents. The Telpherage Company seem in good hopes – and I hear at all hands that the opening was felt to be really successful.'[93] But the project stalled without Fleeming Jenkin as driving force. Despite Ayrton and Perry's key role in the genesis of telpherage, Perry admitted in 1886 'how hopeless it was to compete with [Professor Jenkin] in the practical working out of an idea'. Jenkin 'was felt by everyone to be the soul of the enterprise'.[94]

In 1889 the telpherage patents were sold to the United Electrical Engineering Company, which shortly afterwards merged with J.G. Statter and Co. to become the Electrical Engineering Corporation. The directors of this new company were J.E. Waller and E. Manville, consulting engineers who built a conduit tramway at Northfleet, and a telpher line at East Pool tin mine, near Redruth in Cornwall.[95] These schemes attracted wide publicity, and Waller and Manville were asked to supply a passenger-carrying telpher to the Edinburgh Exhibition of 1890. So ironically the first outdoor telpherage line used for public transport appeared in Jenkin's home city, almost five years to the day after his death, installed not by his son or associates, but by two other, scarcely known electrical engineers.

Whether Jenkin's telpher project was a failure is debatable. Commercially it did not work, for as Jenkin had foreseen it could not compete with steam or other electrical systems in many applications. It is unfair to label it a toy, though – as a scathing article in *The Electrician* did in 1890 – for it was seriously intended and seriously developed.[96] Although Jenkin had worked out telpherage in great detail, when he died it was still a prototype. It is clear – and was clear to Jenkin – that changes would have been necessary in response to problems revealed during the first commercial operation of the line. Perry and others did make alterations during and after 1885, especially in separating

---

[91] Kelvin, Cambridge, E114, Ewing to Thomson, 27 October 1885; Hempstead 1993, p. 203.

[92] Venn (ed.) 1947, p. 560.

[93] Kelvin, Cambridge, J18, Anne Jenkin to Thomson, 8 November 1885.

[94] Hempstead 1993, p. 204.

[95] Bond 1969, p. 23. Jenkin's US patents were the subject of a long legal action, by Rudolph M. Hunter, from 1884. Hunter finally lost the case in 1891: IEE, SC mss 51/3.

[96] Hempstead 1993, p. 203, quoting *The Electrician* XXV (1890), p. 666.

load- from current-bearing functions in the cables, and also introducing human control to the locomotives so that the system never became fully automatic.[97] It is safe to say that Jenkin would have done it differently. His originality, built upon extensive mechanical and electrical experience and a unique detailed understanding of the telpherage scheme, proved impossible to replicate.[98] In any case, telpherage was not a blind alley in the history of technology, for a line of development is traceable from Jenkin's telpherage to the *téléphérique* – modern cable cars and chair lifts.[99]

By the closing years of his life Jenkin was well known, his eminence in his profession widely acknowledged. From an associate of the Institution of Civil Engineers in 1859, he had moved to full membership in 1868. He became a Fellow of the Royal Society in 1865, supported by Thomson, Clerk Maxwell, Fairbairn, and Galton.[100] In 1869 he was elected to the Royal Society of Edinburgh, becoming its Vice President ten years later. He was president of the Mechanical Section of the British Association in 1871 and was elected to the Institution of Mechanical Engineers in 1875, proposed by Edward Easton, William Siemens and his old employer John Penn. He joined the Society of Telegraph Engineers, forerunner of the Institution of Electrical Engineers, soon after it was established in 1871.

Before he was thirty Jenkin had been juror for physical apparatus and reporter for electrical apparatus at the 1862 Exhibition. He acted as juror in engineering at the 1878 Paris Exhibition, taking Stevenson as his secretary, the only paid employment which the author ever held. It is not clear how Stevenson earned his money in Paris, for Ewing noted that while he had many letters from Jenkin during the course of that summer, none was written by the secretary.[101] Jenkin was also juror in electricity at the Health Exhibition the year before his death.

In 1883 he contemplated publishing a collection of his best papers. He had written only two full-length books, the textbooks on electricity, one of which he had also translated – 'The first *Electricity* is now extant in my choice Italian, German and French'. Several of his non-scientific papers he considered to have permanent value: the review of Darwin, Lucretius, the article on population, the laws of supply and demand, two papers on Mrs Siddons, and the review of Agamemnon.[102] The idea of republishing these was not taken up again until after his death, when these and other works accompanied Stevenson's memoir.

---

[97] Hempstead 1994, p. 59.
[98] Hempstead 1993, pp. 203–4.
[99] Hempstead 1998, p. 36.
[100] See the citation in the *Royal Society Printed List of Candidates, 1848–1867*.
[101] Ewing 1933, pp. 269–70.
[102] Kelvin, Cambridge, J31, Jenkin to Thomson, 7 February 1883.

It would be a mistake to see Jenkin's career as in decline by 1885. While much of his energy focused upon telpherage, as ever he taught, he wrote on a range of issues, he was active in societies and associations. Indeed he was described in an obituary as being 'in the very height of his usefulness'.[103] In a few years more, he would have taken his turn as President of the Society of Telegraph Engineers and of the Royal Society of Edinburgh.[104] The University of Glasgow recognised his distinguished work in conferring an honorary doctorate, an LLD., in 1883. Had he lived longer, Jenkin could, like Thomson, have received a knighthood or greater honour to acknowledge the services which he had given, voluntarily and often unpaid, to electrical science, to education, and to public health.

But Jenkin's health was beginning to fail. Since the emergency after Frewen's birth, he had not been strong. He no longer had need to worry about money or future security, and counselled Ewing, who was suffering financial and family anxieties in Japan similar to those which had troubled Jenkin in Claygate, to have courage. 'As regards bread and butter a man like you has simply nothing to fear. Come home as soon as possible.' But ominously he told his former student: 'Live your life gaily. When misfortune comes, suffer like a man, and cast the suffering away as soon as you can; but a life spent in scanning the horizon for conceivable storms is not wisely led. Our will is master of that sort of thing, believe me.'[105] The years may have taught Jenkin the lesson of patience – as his comments on electric lighting suggest – but he had not learned caution. His nature precluded it, and he shrank from delegating responsibilities. 'Being fully convinced that the world will not continue to go round unless I pay it personal attention, I must run away to my work', he had once told his wife.[106] It was only partly a joke.

The year after this seemingly frivolous remark, Jenkin admitted to James Clerk Maxwell that he was paying a price for overwork. He had sent on, unread, a manuscript to Maxwell in July 1874: 'I intended to read it but have been so overwhelmed with work. So much so that I mean to stop everything for the next two months'.[107] At that time Ewing was able to take on much of the telegraphy load, and act as right-hand man in Jenkin's scientific work. The two were close personally and shared a professional empathy. After Ewing's departure to Japan in 1878, Jenkin did not again find such an associate.

---

[103] Obituary, *Institution of Civil Engineers Proceedings*, p. 373.
[104] The Society of Telegraph Engineers (STE) was founded in 1871, becoming the Institution of Electrical Engineers in 1891.
[105] Ewing 1933, p. 256.
[106] RLS, p. cxiii. Jenkin wrote this in 1873.
[107] Cambridge University Library, Add 7655/ II/84, Jenkin to Maxwell, 25 July 1874.

In the winter of 1883, while immersed in the problems of telpherage, Jenkin suffered a 'sharp illness [which] gave warning that he was overtaxing even his power of work'.

He appeared to make a full recovery, 'and fell to again with little if any abatement of zeal'.[108] There were other, domestic, anxieties. Henrietta Jenkin's health had caused her son much worry from the autumn of 1875 when she suffered a stroke which left her without speech or hearing. She was at first determined to recover her faculties and made some improvement, but over the following years stroke followed upon stroke, leaving her permanently disabled and dependent upon others. While retaining some memory and understanding, Jenkin's mother had become a pathetic shadow of her former brilliant self. The Captain, who according to Stevenson had been eclipsed by his wife during most of the marriage, cared for her uncomplainingly, with tenderness and sensitivity. In their last years Charles and Henrietta Jenkin displayed a deep and uncomplicated love which had not been obvious before her illness, and their golden wedding in 1881 was a poignant occasion.[109]

The close family group around Jenkin fragmented within a year after the death of Anne Jenkin's father, Alfred Austin, in May 1884. Her mother Eliza died the following January. Within days of this the Captain fell ill, dying aged eighty-three on 5 February 1885. The news of his death was kept from Mrs Jenkin, who lay bedridden in another room. She survived him by forty-nine hours, her death occurring in the first minutes of her seventy-seventh birthday, 8 February 1885.[110] Jenkin took the bodies of his parents by train from Edinburgh to King's Cross, across London by horse-drawn hearse, and from Charing Cross to Stowting for burial.[111]

After this ordeal, Jenkin contemplated a long break from work. He had been badly shaken by the series of deaths, on top of his fatigue from overworking. Stevenson thought that the telpherage scheme had 'consumed his time, overtaxed his strength and overheated his imagination'.[112] Acknowledging that he was exhausted, he had appointed a secretary, F.J. Woods, by February 1885.[113] To Thomson in April 1885, appended to a business letter written by Woods, Jenkin noted: 'I am going to ask leave of absence from Edinburgh next winter. These two last winters have done me up very much.'[114] That same month, the university senate approved Jenkin's request for leave, on grounds of ill-health. He had sent in a medical certificate and arranged for Professor Henry

[108] Obituary, *Institution of Civil Engineers Proceedings*, p. 372.
[109] RLS, pp. cxlvi–cl.
[110] RLS, p. clii. Stevenson is vague about Mrs Jenkin's age, which is given in *The Scotsman*, 9 February 1885, and confirmed on the gravestone at Stowting.
[111] Folkestone Library, Dunk mss, Undertaker's ledger, f1963/5/B3.
[112] RLS, p. cliii.
[113] Kelvin, Cambridge, J33, Jenkin to Thomson, 9 February 1885.
[114] Kelvin, Cambridge, J39, Jenkin to Thomson, 1 April 1885.

Dyer to carry out all teaching and examination duties during the winter session for a salary of £350, to be paid by Jenkin.[115]

Jenkin's plan was to visit Italy with his wife, their first time alone together since the birth of their children. Through a complete rest from work, and living abroad for a time, he expected to recover his health and energy.[116] At the end of the university session in 1885, he went to London to continue work on the telpherage project.[117]   Early in June Jenkin returned to Edinburgh for minor surgery, a small operation on his foot intended to allow him again to enjoy long walks with Anne on their planned 'real honeymoon tour'.[118] The operation appeared to have been successful, but when Anne was reading to him soon afterwards she noticed that his mind seemed to wander. He died without regaining full consciousness, at his home, on 12 June 1885. He was fifty-two. The cause of death was a chance blood poisoning, septicaemia, the scourge of nineteenth-century surgery. It was a cruel irony that the pioneer of antiseptic surgery Joseph Lister had been an active fellow member of the university senatus with Jenkin, holding the Edinburgh chair of clinical surgery between 1869 and 1877. Lister's system of aseptic surgery, developed between 1865 and 1867, was not adopted in time to save his former colleague.[119]

The long journey was retraced to Stowting, where Fleeming Jenkin was laid in the churchyard near his parents and past generations of Jenkins. There had been no other burials in the tiny parish since February, and his name follows directly after those of Charles and Henrietta in the burial register.[120] On his memorial was engraved: 'Tribulation worketh patience; and patience experience; and experience hope'.

Jenkin's unexpected death brought genuine and heartfelt tributes. The university senate recorded its shock:

> They feel that, as an active and efficient Teacher, he has left a place which will not easily be filled; they remember the distinction which, as a man of practical science who had won for himself a European reputation, he conferred on the University by his tenure of one of its chairs; and they think regretfully of the blank in their own number, and in the Society of

[115] Edinburgh University Archives, Senatus Minutes, VIII, 1883–6, pp. 297–98.
[116] Obituary, *Journal of the Society of Telegraph Engineers*, p. 345.
[117] Obituary, *The Electrician*, 19 June 1885, p. 97.
[118] RLS, p. cliv.
[119] See Rickman John Godlee, *Lord Lister* (Macmillan, 1918), pp. 124, 178. Godlee could write in 1918: 'It is a rare event now for septicaemia to follow a surgical incision through unbroken skin'.
[120] Parish register of St Mary the Virgin, Stowting, Kent, 1885, numbers 334, 335 and 336. This burial register, begun in 1813, is still in use. The Reverend Father David Hancock traced these entries and located memorials to the Jenkin family.

Edinburgh, caused by the removal of a colleague of such bright and stimulating personal qualities and so many fine accomplishments.[121]

Robert Louis Stevenson was stunned by the loss of his friend and mentor. It was quickly agreed that he would write the *Memoir* to accompany a collection of Jenkin's papers, to be edited by Anne Jenkin and Sidney Colvin.[122] In the event Thomson, who was to have advised on publication of the scientific papers, proved less than helpful, and Alfred Ewing took over that role, becoming co-editor with Colvin. Anne Jenkin was instrumental in producing the *Memoir* and collected writings, working closely with Stevenson, collecting information and liaising with publishers. Despite her considerable intellect and talents, the woman who could hold a theatre enraptured with an acting performance, who had been persuaded by her husband to lecture on health to poor women and girls, and who had even seen some of her own work published, had little confidence in her own writing ability and preferred to leave most of the editing to Jenkin's friends.[123]

The sudden death of Jenkin completely and permanently disrupted family life in Great Stuart Street. The house was quickly shut up, and later sold. Anne kept on the lodge at Glenmorven, where Colvin visited for consultations while editing the memoir. For years she led a peripatetic existence, in contrast to the settled and stable life of her childhood and marriage. During the winter of 1885 she took a furnished house in Tavistock Square in London, to be close to her sons, and visited Stevenson in Bournemouth several times to help with preparation of the memoir.[124] Later she stayed in Cheshire and Manchester, near Frewen who was assistant to the Works Manager at the London and North Western Railway Company in Crewe. Anne did not meet Stevenson again after he left England for the last time in 1887, though he urged her to visit him in Samoa – 'only fourteen days' journey from San Francisco'.[125] She remained on friendly terms with Fanny Stevenson, the widow, after Stevenson died in 1894.[126] Although Anne Jenkin never visited the Pacific, she did spend time on the continent, in Italy and Belgium. By 1903 she had finally settled in London,

---

[121] Edinburgh University Archives, Senatus Minutes, VIII, 1883–6, pp. 318–9, 26 June 1885.

[122] Booth and Mehew (eds) 1995, V, p. 117, Stevenson to Anne Jenkin, late June 1885.

[123] Ewing 1933, p. 254. Anne had written an article, 'Highland Crofters', which was published in the Presbyterian journal *Good Words* (1885), pp. 502–8. After Stevenson's death, some of her recollections of him appeared in the *Edinburgh Academy Chronicle*, March 1895, pp. 50–52.

[124] See Kelvin, Cambridge, E114, Ewing to Thomson, 27 October 1885; also J18, Anne Jenkin to Thomson, 8 November 1885.

[125] Booth and Mehew (eds) 1995, VII, pp. 452–3, Stevenson to Anne Jenkin, 5 December 1892.

[126] See National Library of Scotland, ms 9895, ff 128–29, 182–83, 250–51; ms 9896, ff 5–6, 57, 139.

at 12 Campden Hill Square where she spent her remaining years. She died on 26 February 1921, aged eighty-three, a widow for thirty-five years, and having outlived a son and a grandson.[127]

The Jenkin boys had been encouraged in their interest in engineering from an early age, with the house the scene of constant experiments: 'Frewen is deep in parachutes. I beg him not to drop from the top landing in one of his own making'.[128] The eldest boy, Austin, was described as an eight year-old by Stevenson as 'a disgusting, priggish, envious, diabolically clever little specimen'.[129] Stevenson claimed to have feared Austin 'when he was about twelve with a perfectly slavish terror' [130] The terror was aroused by Austin's cleverness, for the boy's mother thought him 'as good and gentle and kind as he is big', his father that he was 'so truly good all through'.[131] Austin went up to Trinity College, Cambridge, in 1880 after the University of Edinburgh, taking a first in mathematics and graduating as twenty-fourth Wrangler. Jenkin told Stevenson that had the examination hall 'not been stiflingly close, he would have been a few places higher'. He added that he and Austin were both quite content with the result – 'We have never supposed he had a genius for mathematics – a scholar of Trinity and a first class man is quite good enough'.[132] Austin afterwards took to the law, the profession of his mother's family, was admitted to the Inner Temple in 1883 and called to the Bar four years later.[133] He married Elizabeth Jane Bowman, an auctioneer's daughter from Cambridge, in 1892, a match evidently disapproved by his mother. As Fleeming Jenkin had advised Stevenson on his marriage, Stevenson in his turn tried to console Anne Jenkin: 'To be sure it is always annoying when people choose their own wives ... Austin would not have dreamed of this if he had not been much in earnest'.[134] Austin Jenkin died in April 1910 at the age of forty-eight and was buried near his father at Stowting

Frewen Jenkin, after a career as a practical engineer which included ten years with Siemens Brothers, was elected first professor of Engineering Science at Oxford in 1908. He had married in 1889 a former neighbour from Great Stuart Street, Mary Oswald Mackenzie (1866–1954), the daughter of a

[127] Frewen's younger son, Conrad, born 1894, a regular officer in the Royal Navy, died in 1916 as the result of an abdominal obstruction: PRO, ADM 196/56, p. 2.
[128] RLS, p. cxviii.
[129] Booth and Mehew (eds) 1994, I, p. 198, Stevenson to his mother, 4 August 1870.
[130] Booth and Mehew (eds) 1995, VIII, p. 117, Stevenson to Anne Jenkin, 18 June 1893.
[131] Ewing 1933, p. 254; RLS, p. cxix.
[132] Yale, Beinecke, 4990, Jenkin to Stevenson, 3 August 1883.
[133] Venn (ed.) 1947, p. 560.
[134] Booth and Mehew (eds) 1995, VII, pp. 452–53, Stevenson to Anne Jenkin, 5 December 1892.

194	A VICTORIAN ENGINEER: FLEEMING JENKIN

judge.[135] At the outbreak of war in 1914 Frewen joined the Navy, and on the formation of the Royal Air Force became head of the branch responsible for aircraft materials, with the rank of lieutenant colonel. This led into scientific research on materials fatigue which continued at Oxford and after his retirement from the chair in 1929. He died at St Albans in 1940.[136] The youngest son, Bernard Maxwell Jenkin, an engineering student at Edinburgh at the time of his father's death, was the only one of the brothers not to go on to Cambridge. He followed a practical engineering career, establishing the engineering company of Kennedy and Jenkin, later Kennedy and Donkin, and lived until 1951.[137]

It fell to Austin Jenkin, the lawyer and only son then of age, to wind up Jenkin's affairs. He had been appointed trustee alongside his mother and Alfred Ewing. It was a complicated matter, valuing assets in patents, copyrights, partnerships and companies, besides settling with personal and professional creditors. In total Jenkin's personal estate – revised after a number of other assets came to light – amounted to almost £20,000. In addition he owned real estate, including the Stowting Court Estate, Mill Farm, land and manorial rights there, a farm in Great Bentley, Essex and others in Northiam and elsewhere in Kent.[138] Payments were still arriving from the French Atlantic company and others indebted to the Thomson, Varley and Jenkin partnership, and Austin Jenkin struggled to produce a final balance sheet.[139] Matters were made more difficult because of Varley's recent death, in 1883.[140] The very informality which Jenkin had urged upon Thomson and Varley, as a way of saving pointless argument in order to use time most efficiently on their individual researches, ultimately caused his son some difficulty. Yet the effectiveness of the partners' unceremonious practices is clear from their financial success. In 1886–87, the first full year following Jenkin's death and more than three years after that of Varley, the partnership still recorded a surplus of over £1100.[141]

In view of the very profitable relationship which Thomson had enjoyed with Jenkin – besides the financial gain, Thomson had benefited from their

[135] Booth and Mehew (eds) 1995, VI, p. 362, n.

[136] An obituary of Charles Frewen Jenkin appears in *The Times*, 26 August 1940.

[137] 'Bernard Maxwell Jenkin, 1867–1951', supplement to the *Kennedy and Donkin House Journal*, June 1952.

[138] Scottish Record Office, SC70/1/244, pp. 802–24; SC70/1/247, pp. 351–57; SC70/4/214, pp. 881–910.

[139] Kelvin, Cambridge, J19, Austin Jenkin to Thomson, 12 March [1887].

[140] Cromwell Fleetwood Varley, who like Jenkin came from a respectable but relatively impecunious background, left £46,000: Jeffery 1997, p. 274. In the final few years of his life, Varley had retired from engineering and devoted most of his time to investigating psychic phenomena: thanks to John Varley Jeffery for this information.

[141] Kelvin, Cambridge, J21, Austin Jenkin to Thomson, 27 October 1887.

scientific and technical collaboration and was relieved of commercial burdens because Jenkin had been prepared to shoulder so much responsibility – in view of this, there is something strange about Thomson's reaction to the death of a man who at times was more than a colleague, whom he apparently considered a friend. For one thing, Thomson's tributes to Jenkin were less than fulsome. An unpleasant section in his note on Jenkin's contribution to science, reproduced as Appendix I by Colvin and Ewing, was excised by Colvin, who wrote to Stevenson that two concluding paragraphs were to be cut out: 'Left standing, their chilly tone of patronage could not but shock.'[142] Thomson had also made plain his lack of interest in helping with the publication of Jenkin's papers, failing to answer Anne Jenkin's queries or meet deadlines, forcing his replacement as scientific editor by Ewing.[143] The obituary notice of Jenkin which Thomson wrote for the Royal Society is laced with references to the influence of Thomson himself upon Jenkin's career. This appears to have been Thomson's consistent view, that any importance which attached to Jenkin flowed only from his relationship with Thomson, a relationship in which Jenkin was very much subordinate.[144] There is nowhere the sense of shock and grief which Thomson expressed upon the sudden death of Sir William Siemens in November 1883.[145] Jenkin's loss does not seem to have touched Thomson in the way that it affected another long-standing associate, Henry C. Forde, who wrote with true regret of the death of 'poor Jenkin'.[146] Jenkin was not a man easily replaced, yet ways had to be found, and Thomson made new arrangements with Forde and Latimer Clark to carry on their work together.[147]

Jenkin had gone, was sincerely regretted and long mourned by many. His sudden death was a blow to his friends, expressed more candidly in private letters than in formal obituary notices. A 'great store of pleasure has gone out of my life', Stevenson wrote to his mother.[148] Jenkin was admirable, upstanding, energetic, useful. He combined these essential qualities of Victorian rectitude with something more, the spark of an enquiring mind, an enduring childlike wonder with everything around him, a happy disposition. The story of his life is more than a window upon a long-dead world of engineers and academics and gentlemen scholars, for Jenkin connected, he communicated, he influenced. He was present at the birth of electrical engineering and was largely responsible for codifying it and for marrying its

---

[142] Yale, Beinecke, 4390, Colvin to Stevenson, 13 September [1887].

[143] This is clear from letters to Thomson from Anne Jenkin, in Kelvin, Cambridge, J14, 4 March [1886]; J15, 18 December [1885]; J17, 27 December [1885]; J18, 8 November [1885].

[144] Susan Morris makes this point strongly: 1994, pp. 322–23.

[145] Kelvin, Cambridge, LB4.260, Thomson to Jenkin, 20 November 1883.

[146] Kelvin, Cambridge, Add 7342, F253, Forde to Thomson, 14 December 1885.

[147] Kelvin, Cambridge, Add 7342, F254, Forde to Thomson, 21 January 1886.

[148] Booth and Mehew (eds) 1995, V, p. 132, mid-October 1885.

theoretical and practical aspects. Furthermore, like Mill he saw all human knowledge, physical or social, as a type of science.[149] His achievements grew from his ability as an intellectual to apply widely the practical experience and techniques learned from his own subject. The marriage of his two devotions, art and science, which brought all that satisfied, was a true marriage. The two were one, for how could engineering be understood without poetry?

[149] Deane 1978, p. 87.

Bibliographical Notes on the History of Submarine Telegraphy

From the time of the earliest attempt to cross the channel with a submarine cable, periodicals such as the *Illustrated London News* attempted to satisfy the intense public interest which the expeditions generated. Lengthy accounts of voyages were accompanied by engravings of cable-making and cable-laying, of ships, sailors, electricians and businessmen. The first books about the Atlantic telegraph appeared before the line itself was in operation. The best known of these is the heroic narrative by W.H. Russell, the *Times* correspondent famous for his reports on the Crimean War, who was on board the Great Eastern for the 1865 attempt on the Atlantic. Although unable to report a successful outcome, Russell provided a graphic and largely accurate, if hagiographic, account of the process of laying undersea telegraphs.

Once submarine cable-laying had become a routine activity, there was an outpouring of books and articles about the birth of long-distance telegraphy. Many of these aimed to explain or justify the actions of various participants – examples are the works of Henry Field, concerned to assure the place of his brother Cyrus Field at the centre of the story of Atlantic telegraphy; and the writings of R.S. Newall, staking his own claim as an innovator in submarine cables. George Saward, secretary to the Atlantic Telegraph Company, wrote his first account in 1861 in the hope of encouraging investment in a renewed attempt upon the Atlantic, and his later book was at least partly a defence of his own and his company's business dealings.

The most thorough and authoritative nineteenth-century work on the subject is that by Charles Bright, nephew of the engineer of the same name who had been knighted at the age of twenty-six for his part in laying the Atlantic cables of the 1850s. When his nephew wrote in 1898, the intercontinental telegraph network was almost complete, waiting only the completion of the Pacific cable, so that it was then possible to present a comprehensive historical and contemporary study of submarine telegraphy. Bright ranged widely across scientific, engineering and social issues, presenting a clear account of the development and practice of submarine telegraphy, and his book remains an invaluable source of reference.

Following the pattern set by Bright, several twentieth-century authors have produced popular accounts which focus upon technological development in undersea cables. Among the most authoritative and readable are those by Bern Dibner and by Bernard S. Finn. Finn's Science Museum booklet remains the best introduction to the history of submarine telegraphy, replete with illustrations and drawing a sharp picture of the technology, although more recent scholarship challenges some of his conclusions. Haigh's book on cable ships also provides formidable detail about cable and telegraph companies and the individuals involved in their promotion.

Biographies of some of the leading nineteenth-century electrical scientists and engineers range from memoirs by contemporary admirers – Pole on Siemens, Thompson on Thomson – to more recent scholarly works, such as Smith and Wise on Thomson, and Goldman on Clerk Maxwell. While Maxwell's theory of electromagnetism is not readily accessible to most readers, Goldman's work gives a gentle entry into the life of a complex man, his motivations and his work.

Similarly, company histories range from earlier publications in a less critical style, to more incisive modern works. There are studies of telegraph companies themselves (Baglehole, Barty-King); of cable-makers and electrical engineers (Scott, Lawford and Nicholson); and of other companies which played significant roles in the development of submarine telegraphy, notably Read's updated study of Reuters. There are also official histories of the Institution of Electrical Engineers, by Appleyard and by Reader.

More recently, a range of books and articles have considered single issues relating to nineteenth-century undersea cables, including their impact on commerce, culture and international relations. Headrick, for example, considers how telecommunications became an invaluable arm of government, made to serve the needs of international diplomacy. Hunt has written extensively on submarine telegraphs in the context of the history of electricity. His book *The Maxwellians* is an account of personal relationships and differences involved in the foundation and growth of electromagnetic field theory, a peculiarly British theory which owed its origins and development to the British interest in submarine telegraphs. Other papers by Hunt are first class studies of the impetus given to the development of electrical theory and instruments by the technical and scientific demands of deep-sea cables.

Papers to the 'Semaphores to Short Waves' conference at the Royal Society of Arts in July 1996 have been collected by Frank James in a book of the same title, which has several contributions on the development and effects of submarine telegraphs.

Others have added substantially to the historiography of submarine telegraphy, and continue to write on the subject. Donard de Cogan has taken a wide view of submarine telegraphs and illuminates scientific, operational and political aspects, particularly of transatlantic cables. Colin Hempstead is interested in the technological and social significance of cables, in particular their contribution to scientific knowledge and their role in driving scientists and engineers to design and construct a host of instruments for measurement and operational purposes.

There are also some significant unpublished dissertations, such as those by Cain, Jaras and Sloan, which cover important elements of the subject. But as yet submarine telegraphy has not found its Boswell, and there is crying need for a comprehensive study which would consider technological developments in tandem with economic and political influences and effects.

References

Rollo Appleyard, *The History of the Institution of Electrical Engineers*, 1871–1931 (London: IEE, 1939).

K.C. Baglehole, *A Century of Service: a Brief History of Cable and Wireless*, 1868–1968, (London: Cable and Wireless, 1969).

Hugh Barty-King, *Girdle Round the Earth: the Story of Cable and Wireless and its Predecessors*, (London: Heinemann, 1979).

Charles Bright, *Submarine Telegraphs: their History, Construction and Working*, (London: Crosby Lockwood, 1898).

Robert J. Cain, 'Telegraph Cables in the British Empire, 1850–1900', unpublished Ph.D. thesis, Duke University, 1971.

Donard de Cogan, 'Dr E.O.W. Whitehouse and the 1858 Transatlantic Cable', *History of Technology*, X (1985), pp. 10–15.

Donard de Cogan, 'Cable Landings in and around Newfoundland', *Aspects*, XXVII (1992), pp. 25–36.

Donard de Cogan, 'Cable Talk: Relations between Heart's Content and Valentia Cable Stations, 1866–86', *Aspects*, XXVIII (1993), pp. 37–43.

Donard de Cogan, 'Ireland, Telecommunications and International Politics', *History Ireland*, Summer 1993, pp. 34–8.

Bern Dibner, *The Atlantic Cable*, (Norwalk, Conn: Burndy Library, 1959).

Henry M. Field, *The Story of the Atlantic Telegraph*, (New York: Charles Scribner's Sons, 1893).

Bernard S. Finn, *Submarine Telegraphy: The Grand Victorian Technology*, (London: Science Museum, 1973).

M. Goldman, *The Demon in the Æther*, (Edinburgh: Paul Harris, 1983).

Kenneth R. Haigh, *Cable Ships and Submarine Cables*, (1968).

Daniel R. Headrick, *The Invisible Weapon: Telecommunications and International Politics*, 1851–1945 (Oxford: Oxford University Press, 1991).

Colin A. Hempstead, 'Kelvin, Instrumentation and the First Atlantic Telegraphy', *Proceedings 14th Annual Meeting of the History of Technology Group, Institution of Electrical Engineers*, (London: Institution of Electrical Engineers, 1987).

Colin A. Hempstead, 'Automatic Senders: Science, Technologies or Art?', *Proceedings 24th Annual Meeting of the History of Technology Group, IEE*, (London: Institution of Electrical Engineers , 1997).

Colin A. Hempstead, 'The Early Years of Oceanic Telegraphy: Technology, Science and Politics', *Institution of Electrical Engineers Proceedings*, 136A, (1989), pp. 297–305.

Colin A. Hempstead, 'Representations of Transatlantic Telegraphy', *Engineering Education and Science Journal*, IV (1995), pp. S17–25.

Bruce J. Hunt, '"Practice v Theory": the British Electrical Debate, 1888–91', *Isis*, LXXIV (1983), pp. 341–55.

Bruce J. Hunt, 'Michael Faraday, Cable Telegraphy and the Rise of Field Theory', *History of Technology*, XIII (1991), pp. 1–19.

Bruce J. Hunt, *The Maxwellians* (Ithaca, N.Y: Cornell University Press, 1991).

Bruce J. Hunt, 'The Ohm is where the Art is: British Telegraph Engineers and the Development of Electrical Standards', *Osiris*, IX (1994), pp. 48–64.

Bruce J. Hunt, 'Scientists, engineers and Wildman Whitehouse: measurement and credibility in early cable telegraphy', *British Journal for the History of Science*, XXIX (1996), pp. 155–69.

Bruce J. Hunt, 'Doing Science in a Global Empire: Cable Telegraphy and Electrical Physics in Victorian Britain', in B. Lightman (ed.), *Victorian Science in Context* (University of Chicago Press, 1997).

Frank A.J.L. James (ed.), *Semaphores to Short Waves* (London: Royal Society of Arts, 1998).

Thomas F. Jaras, 'Promoters and Public Servants: the Role of the British Government in the Expansion of Submarine Telegraphy, 1860-1870', unpublished Ph.D. thesis, Georgetown University, 1975.

G.L. Lawford and L.R. Nicholson (eds), *The Telcon story, 1850–1950* (London: Telegraph Construction Company, 1950).

R.S. Newall, *Facts and Observations relating to the invention of the submarine cable and to the manufacture and laying of the first cable between Dover and Calais* (London: E and F.N. Spon, 1882).

William Pole, *The Life of Sir William Siemens* (London: John Murray, 1888).

Donald Read, *The Power of News: the History of Reuters* (Oxford University Press, 1999).

W.J. Reader, *A History of the Institution of Electrical Engineers, 1871-1971* (London: Peter Peregrinus, 1987).

William H. Russell, *The Atlantic Telegraph* (London: Day and Son, 1865).

George Saward, *Deep-Sea Telegraphs: their Past History and Future Progress* (London: Mechanics' Magazine, 1861).

George Saward, *The Trans-Atlantic Submarine Telegraph: a brief Narrative of the Principal Incidents in the History of the Atlantic Telegraph Company* (London: privately published, 1878)

John D. Scott, *Siemens Brothers, 1858–1958* (London: Weidenfield and Nicolson, 1958).

Norton Q. Sloan, 'William Thomson's inventions for the submarine telegraph industry: a 19th century technology program', unpublished Master of Liberal Arts thesis, Harvard University, 1996.

Crosbie Smith and M. Norton Wise, *Energy and Empire: a Biographical Study of Lord Kelvin* (Cambridge University Press, 1989).

Silvanus P. Thompson, *The Life of William Thomson, Baron Kelvin of Largs* (London: Macmillan, 1910), 2 vols.

Bibliography

Published and Unpublished Works of Fleeming Jenkin

'On Gutta Percha as an Insulator at Various Temperatures', *British Association for the Advancement of Science Report* (1859), p. 248; also in appendix to the Galton Report; abstract in *Civil Eng. & Archit. Journ.* 22 (1859), pp. 321–22.

'On the Retardation of Signals through long Submarine Cables', *British Association for the Advancement of Science Report* (1859), p. 251; also in 'Experimental Researches ...' below, 1862.

'On the Insulating Properties of Gutta-Percha', *Proceedings of the Royal Society*, X (1860), p.409; also in appendix to the Galton Report; also *Philosophical Magazine*, ser. IV, 21 (1861), pp. 75–79.

Rapport de M. Jenkin, ingenieur electrician charge de la reparation du cable entre Cagliari et Bone en 1860 (Cagliari, 1860)

'On Permanent Thermoelectric Currents in Circuits of One Metal', *British Association for the Advancement of Science Report* (1861), p. 39; also *Chemical News*, IV, 26 Oct 1861, p. 222.

'On the True and False Discharge of a Coiled Electric Cable', with William Thomson, *Philosophical Magazine*, ser. IV, 22 (1861), pp. 202–11; also Article LXXXIII of Thomson's *Mathematical and Physical Papers*, II.

'Experimental Researches on the Transmission of Electric Signals thro' Submarine Cables – Part I. Laws of Transmission thro' various lengths of one cable', *Philosophical Transactions*, 152 (1862), pp. 987–1017; abstract in *Proceedings of the Royal Society*, XII (1862), p. 198.

'Appendix H to the First Report of the Committee on Electrical Standards', *British Association for the Advancement of Science Report* (1862), p. 125.

Report ... [on] a telegraphic cable to be laid by the Indian government in the Persian Gulf, with H.C. Forde (London: Spottiswoode, 1862).

'Description of the Electrical Apparatus arranged.. for the Production of Exact Copies of the Standard of Resistance', *British Association for the Advancement of Science Report* (1862), p. 159.

'On Thermo-electric Currents in Circuits of one Metal', *British Association for the Advancement of Science Report* (1862), p. 173.

'On the Construction of Submarine Telegraph Cables', *Minutes of the Proceedings of the Institution of Mechanical Engineers* (July 1862), pp. 211–41.

'Report on Electrical Instruments' in *Jurors' Reports for the International Exhibition, London, 1862* (London, 1863); also in *Annales Telegraph*, VII (1864) and VIII (1865).

'Appendix C to the Second Report of the Committee on Electrical Standard', with J. Clerk Maxwell, *British Association for the Advancement of Science Report* (1863), p. 111; also *Philosophical Magazine*, XXIX (1865).

'Appendix D to the Second Report of the Committee on Electrical Standards', with J. Clerk
   Maxwell and Balfour Stewart, *British Association for the Advancement of Science
   Report* (1863).
'On the Elementary Relations between Electrical Measurements', *British Association for the
   Advancement of Science Report* (1863), p. 130.
'Description of an Experimental Measurement of Electrical Resistance', *British Association for
   the Advancement of Science Report* (1863), p. 163.
'The Origin of Species', *North British. Review*, XLVI (OS), VII (NS), June 1867, pp. 277–318;
   review of Darwin's *Origin of Species*, also in Colvin and Ewing (eds) 1887, I, as
   'Darwin and the Origin of Species'.
Lectures on the Construction of Telegraph Lines (Royal Engineers Institute, Chatham, 1863),
   18pp. (Written from Jenkin's notes by C.S. Beauchamp, and corrected by Jenkin).
Lectures on the Maintenance and Efficiency of Telegraphic Lines (Royal Engineers Institute,
   Chatham, 1863).
'Description of a further Experimental Measurement of Electrical Resistance ... Appendix A to
   the Third Report of the Committee on Electrical Standards', with J. Clerk Maxwell, also
   Charles Hockin, *British Association for the Advancement of Science Report*, XXXIV
   (1864), pp. 345–51.
'On the Retardation of Electrical Signals on Land Lines', *British Association for the
   Advancement of Science Report*, XXXIV (1864), pp. 13–14; also *Philosophical
   Magazine*, ser. IV, XXIX (1865), pp. 409–21.
'On an Electric-resistance Balance constructed by Prof. W. Thomson', *British Association for
   the Advancement of Science Report* (1864), p. 14.
'Report on the new Unit of Electrical Resistance proposed and issued ...', *Proceedings of the
   Royal Society*, XIV (1865), pp. 154–64; also *Philosophical Magazine*, XXIX (1865),
   pp. 477–86; *Annals of Physics and Chemistry*, CXXVI (1865), pp. 369–87.
'Determination d'un nouvel étalon de résistance électrique', *Annales Telegraph*, VIII (1865),
   pp. 273–81.
'On the Elementary Relations between Electrical Measurements', with J. Clerk Maxwell,
   Philosophical Magazine, ser. IV, XXIX (1865), pp. 436–60, 507–25.
'Report on Standards of Electrical Resistance', *British Association for the Advancement of
   Science Report* (1865), p. 308.
'Letter on Resistance Standard', *Philosophical Magazine*, ser. IV, XXIX (1865), p. 248; also in
   Popular Science Review, IV (1865), p. 397.
'Cantor Lectures on Submarine Telegraphy, Delivered before the Royal Society of Arts,
   January and February 1866', *Journal of the Society of Arts*, XIV (1866), pp. 174–76;
   193–98; 217–21; 233–9; 259–65.
'On a new Arrangement for Picking up Submarine Cables', *British Association for the
   Advancement of Science Report*, (1866), p. 145.
'Submarine Telegraphy', *North British Review*, XLV (OS), VI (NS), December 1866, pp. 459–
   505; also in Colvin and Ewing (eds) 1887, II.
'Reply to Dr. Werner Siemens Paper 'On the Question of the Unit of Electrical Resistance'',
   Philosophical Magazine, ser. IV, XXXII (1866), pp. 161–77; also *Annales de Chimie*,
   X (1867), pp. 92–106.
'On a Modification of Siemens's Resistance Measurer: App. II to 5th Report of the Committee
   on Electrical Standards', *British Association for the Advancement of Science Report*,
   XXXVII (1867), pp. 481–3.
'Experiments on Capacity: App. IV to 5th Report of the Committee on Electrical Standards',
   British Association for the Advancement of Science Report, XXXVII (1867), pp. 483–8.

Lectures on Electrical Measurements at Chatham, February 1867 (Royal Engineer
   Establishment, 1867), 70pp. (Notes of lectures reported by Capt. R.H. Stotherd, RE,
   revised by Jenkin.)
'Review of 'Fecundity, Fertility and Sterility' by Dr Matthews Duncan', *North British Review*,
   XLVII (OS), VIII (NS), December 1867, pp. 441–62.
'The Atomic Theory of Lucretius', *North British Review*, XLVIII (OS), IX (NS), March 1868,
   pp. 211–42; review of the second edition of H.A.J. Munro's *Lucretius*, also in Colvin
   and Ewing (eds) 1887, I.
*A Lecture on the Education of Civil and Mechanical Engineers in Great Britain and Abroad:
   Public Inaugural Address, 3 November 1868* (Edinburgh, 1868).
'On the Rate of Increase of Underground Temperature', *British Association for the
   Advancement of Science Report* (1868), p. 510.
'Trade Unions: How far Legitimate?', *North British Review*, XLVIII (OS), IX (NS), March
   1868, pp. 1–62; review of Reports of Commissioners appointed to enquire into trade
   unions, also in Colvin and Ewing (eds) 1887, II.
*On Technical Education: An Address read before the Royal Scottish Society of Arts in
   Edinburgh, 1869* (Edinburgh,1869); also in Colvin and Ewing (eds) 1887, II.
'On the Practical Application of Reciprocal Figures to the Calculation of Strains on
   Framework', *Transactions of the Royal Society of Edinburgh*, XXV (1869), pp. 441–8.
'French Atlantic Telegraph: Engineers' Final Reports and Appendix', with Latimer Clark and
   H.C. Forde, 17 and 25 September 1869, 36pp., unpublished ms. in New York Public
   Library, Wheeler collection, 4608.
'Report on Standards of Electrical Resistance', *British Association for the Advancement of
   Science Report* (1869), p. 434.
'On the Rate of Increase of Underground Temperature', *British Association for the
   Advancement of Science Report* (1869), p. 176.
'On the Provision existing in the United Kingdom for the vigorous Prosecution of Physical
   Research', *British Association for the Advancement of Science Report* (1869), p. 213.
'On the Submersion and Recovery of Submarine Cables', *Proceedings of the Royal Institution*,
   V (1869), pp. 574–80.
'Note on the Electrification of an Island', *Nature*, II, 5 May 1870, p. 12; also *American Journal
   of Science*, L (1870), pp. 148–9.
'The Graphic Representation of the Laws of Supply and Demand, and their Application to
   Labour', in *Recess Studies*, ed. Sir Alexander Grant (Edinburgh, 1870); also in Colvin
   and Ewing (eds) 1887, II.
'Report on Standards of Electrical Resistance', *British Association for the Advancement of
   Science Report* (1870), p. 14.
'On Braced Arches and Suspension Bridges', *Transactions of the Royal Scottish Society of the
   Arts*, VIII (1870).
'Presidential Address to the Mechanical Science Section of the British Association', *British
   Association for the Advancement of Science Report*, XLI (1871), pp. 225–9.
'On the Principles which regulate the Incidence of Taxes', *Proceedings of the Royal Society of
   Edinburgh*, VII (1872), pp. 618–31; also in Colvin and Ewing (eds) 1887, II.
'Atlantic Telegraphy', *Journal of the Society of Arts*, XXI (1872), p. 362.
'On the Construction and Submersion of Submarine Telegraph Cables', *Journal of the Society
   of Telegraph Engineers*, I (1872–73), pp. 114–23.
Electricity and Magnetism (London: Longmans Textbooks of Science, 1873)
'On Dynamical and Electrical Units', *British Association for the Advancement of Science
   Report* (1873), p. 222.
'On Science Lectures and Organisation', *British Association for the Advancement of Science
   Report* (1873), pp. 495, 507.

'On a Method of Testing Short Lengths of Highly Insulated Wire in Submarine Cables', *Journal of the Society of Telegraph Engineers*, II (1873), pp. 169–75.

First Report of the Committee for the Selection and Nomenclature of Dynamical and Electrical Units (London, 1874).

'On the Antique Greek Dress for Women', *Art Journal*, January 1874; also in Colvin and Ewing (eds) 1887, I.

'On Mr Siemens's Pyrometer', *British Association for the Advancement of Science Report* (1874), p. 242.

'On Dynamical and Electrical Units', *British Association for the Advancement of Science Report* (1874), p. 255.

Report on Hooper's India-Rubber Core, with Thomson, Frankland etc., (London, 1875), 9 pp.

'On a Constant Flow Valve for Water and Gas', *Transactions of the Royal Scottish Society of Arts*, IX, March 1876, pp. 370–76.

Bridges: an Elementary Treatise on their Construction and History (Edinburgh: Black, 1876); reprinted from the *Encyclopaedia Britannica*.

'On Friction between Surfaces moving at Low Speed', with J.A. Ewing, *Philosophical Transactions*, CLXVII (1877), pp. 509–28; also *Proceedings of the Royal Society*, XXVI (1878), pp. 93–4; *Philosophical Magazine*, IV (1877), pp. 308–10; *Journal de Physique*, VI (1877), pp. 285–87.

'On the Application of Graphic Methods to the Determination of the Efficiency of Machinery', *Transactions of the Royal Society of Edinburgh*, XXVIII (1877), pp. 1–36, 703–15; also in Colvin and Ewing (eds) 1887, II; *Association Française pour l'avancement des Sciences, Congrès de Rheims*, August 1880.

'On the Application of Graphic Methods to the Determination of the Efficiency of Machinery: Part II, The Horizontal Steam Engine', *Transactions of the Royal Society of Edinburgh*, XXVIII (1878), p. 703; also in Colvin and Ewing (eds) 1887, II.

'Remarks on the Phonograph', with J.A. Ewing, *Proceedings of the Royal Society of Edinburgh*, IX (1878), pp. 579–81.

'On the Wave Forms of the Vowel Sounds produced by the Apparatus exhibited by Prof. Crum Brown', with J.A. Ewing, *Proceedings of the Royal Society of Edinburgh*, IX (1878), pp. 723–25.

Healthy Houses (Edinburgh: David Douglas, 1878).

'Mrs Siddons as Lady Macbeth', *Nineteenth Century*, (1878); also in Colvin and Ewing (eds) 1887, I; and Matthews (ed.) 1958.

Objects of the Sanitary Protection Association (Edinburgh: SPA, 1878).

'Browning's 'Agamemnon' and Campbell's 'Trachiniae'', *Edinburgh Review*, CXLVII, April 1878, pp. 409–36; review of Browning's *Agamemnon*; Morshead's translation of *The Agamemnon of Aeschylus*; and Lewis Campbell's translation of *Three Plays of Sophocles*, also in Colvin and Ewing (eds) 1887, I.

'On the Harmonic Analysis of Certain Vowel Sounds', with J.A. Ewing, *Transactions of the Royal Society of Edinburgh*, XXVIII (1878), pp. 745–77.

'On the Wave Forms of Articulate Sounds', with J.A. Ewing, *Proceedings of the Royal Society of Edinburgh*, IX (1878), pp. 582–8, 714–18.

'The Phonograph and Vowel Theories', with J.A. Ewing, *Nature*, XVIII (1878), pp. 167–69.

'The Phonograph and Vowel Sounds', with J.A. Ewing, *Nature*, XVIII (1878), pp. 340–43, 394–7, 454–56.

A Lecture on the Education of Civil and Mechanical Engineers in Great Britain and Abroad (Edinburgh, 1878).

'The Time-Labour System, or How to Avoid the Evils caused by Strikes', previously unpublished ms., 1879–81, in Colvin and Ewing (eds) 1887, II.

'Scenes from the Agamemnon ... arranged for the modern stage by Fleeming Jenkin' unpublished ms., 1880.

What is the Best Mode of Amending the Present Laws with Reference to Existing Buildings ... (London: Spottiswoode, 1880); paper read at Congress of Social Science Association, Edinburgh, October 1880.

Electricity (London: SPCK Manuals of Elementary Science, 1881), 128pp.

'Interim Report for Constructing and Issuing Practical Standards for use in Electrical Measurements', *British Association for the Advancement of Science Report* (1881), pp. 1–17.

'Care of the Body', in *Health Lectures for the People* (Edinburgh, 1881); lecture delivered at the Watt Institution.

'On Standards for use in Electrical Measurements', *British Association for the Advancement of Science Report* (1881), p. 423.

'Mrs Siddons as Queen Katharine, Mrs Beverley and Lady Randolph, from contemporary notes by George Joseph Bell', Macmillan's Magazine, XLVI, May 1882, pp. 20–30; also in Colvin and Ewing (eds) 1887, I; and Matthews (ed.) 1958.

House Inspection (Edinburgh, 1882); pamphlet reprinted from the *Sanitary Record*.

Griselda, unpublished play, privately printed for performance in January 1882; in Colvin and Ewing (eds) 1887, I.

'On Standards for use in Electrical Measurements', *British Association for the Advancement of Science Report* (1882), p. 70.

Conference Internationale pour la Determination des Unités Électriques, 16–26 October 1882 (Paris, 1882–84), 161pp. + 117pp. Includes Professors Jenkin, Thomson and D.E. Hughes.

'On Rhythm in English Verse', *Saturday Review*, February/ March 1883 (three articles); review of Edwin Guest, *A History of English Rhythms,* also in Colvin and Ewing (eds) 1887, I.

'Nest Gearing', *British Association for the Advancement of Science Report* (1883), pp. 387–90.

'Telpherage', *The Electrician*, 13, 3 November 1883; introductory address to the class of engineering at Edinburgh, 30.10.83; also in *La Lumière Électrique*, 10 November 1883.

Report on behalf of certain Local Authorities in respect of ... Electric Lighting Act 1882 (Westminster, 1883), 26pp.; relates to lighting of Great Yarmouth.

'Talma on the Actor's Art', *Saturday Review* (1883); review of *Talma on the Actor's Art*, with a preface by Sir Henry Irving, also in Colvin and Ewing (eds) 1887, I; and Matthews (ed.) 1958.

'On Standards for use in Electrical Measurements', *British Association for the Advancement of Science Report* (1883), p. 41.

'Lectures on Heat in its Mechanical Applications, including Gas and Caloric Engines', Proc. Inst. Civil Eng. (1885); a series of lectures delivered to the Institution during the 1883–84 session.

'On Electric Lighting', *The Electrician*, XI (1883), pp. 114–18.

Telpherage (School of Military Engineering, Chatham, 1884).

'Telpherage', paper read to the Society of Arts, 16 May 1884; Colvin and Ewing (eds) 1887, II.

'Is one Man's Gain another Man's Loss?', previously unpublished ms., 1884, in Colvin and Ewing (eds) 1887, II.

'On Science Teaching in Laboratories', paper read at the International Health Exhibition Conference, London 1884; in Colvin and Ewing (eds) 1887, II.

'Telpherage', *Good Words* (1884), pp. 132–39.

Circular to Electric Lighting Exhibitors (London, 1884).

'Artist and Critic', *Saturday Review*, 11 October 1884; also in Colvin and Ewing (eds) 1887, I.

'On Standards for use in Electrical Measurements', *British Association for the Advancement of
Science Report* (1884).
Report on the Telpherage (London: the Telpherage Company, ?1884), 4pp.
'A Fragment on Truth', unpublished and unfinished ms, 1885; in Colvin and Ewing (eds) 1887, I.
Houses of the Poor: their Sanitary Arrangement (1885).
'A Fragment on George Eliot', unpublished and unfinished review of a biography of Eliot by
her second husband, John Walter Cross, 1885; in Colvin and Ewing (eds) 1887, I.

Fleeming Jenkin's Patents

1860 'Telegraphic Communications', with William Thomson. UK number 2047.
1861 'Bridges'. UK number 667.
1863 'Electric Tell Tale Compass'. UK number 1553.
1865 'Winding in Telegraph Cables'. UK number 1218.
1865 'Machinery for Manufacturing Telegraph Cables'. UK number 2155.
1869 'Apparatus for Producing Electric Light'. UK number 390.
1869 'Bridges'. UK number 3071.
1869 'Submarine Telegraph Cables'. UK number 3236.
1873 'V-Pulleys for the Transmission of Power', with F.H. Ricketts. UK number 1886.
1873 'Telegraphic Apparatus', with Sir William Thompson. UK number 2086.
1874 'Obtaining Motive Power'. UK number 2441.
1876 'Telegraphic Apparatus', with Sir William Thomson. UK number 1095. (Automatic
     sender)
1877 'Transmitting Sounds by Electricity', with J.A. Ewing. UK number 4402.
1881 'Caloric Motor Engines', with A.C. Jamieson. UK number 1078.
1881 'Caloric Motor Engines', with A.C. Jamieson. UK number 1130.
1881 'Caloric Motor Engines', with A.C. Jamieson. UK number 1160.
1882 'Mechanism for Transporting Goods and Passengers by Means of Electricity'. UK
     number 1830. (First telpherage patent).
1882 'Regulating Speed in Machinery Driven by Electricity'. UK number 3007.
1882 'Mechanism for Transporting Goods and Passengers by Means of Electricity'. UK
     number 4548.
1883 'Machinery for Spinning and Winding', with J.A. Ewing. UK number 26.
1883 'Driving Gear'. UK number 1913.
1883 'Gearing'. UK number 4481.
1883 'Driving gear called 'Nest Gear''. UK number 4574.
1884 'Gas Engines'. UK number 2635.
1884 'Telpherage'. UK number 3795.
1884 'Trucks and Locomotives for Telpher Lines'. UK number 3796.
1884 'Contact Arms for Telpher Trains'. UK number 4167.
1884 'Telpherage'. UK number 5020.
1884 'Transporting Goods and Passengers by Means of Electricity', with A.C. Elliot. UK
     number 8460.
1884 'Telpher Locomotive'. UK number 8751.
1884 'Regulation of Currents in Telpher and other Electric Motors'. UK number 8906.
1884 'Underground Electric Haulage'. UK number 10907.
1884 'Water Motors', with H. Darwin. UK number 11038.
1884 'Mechanism for the Transmission of Power', with J.A. Ewing. UK number 12479.

1884 'Governors'. UK number 15111.
1884 'Mechanism for the Electrical Transportation of Goods'. USA number 305194.
1886 'Mechanism for Transporting Goods by Electricity'. USA number 343319.

Primary Sources

British Library: Add. Mss 38946 f. 57; 35789 f. 163.
Cambridge University Library: Maxwell letters, Add. 7655; Kelvin collection, Add. 7342.
The Edinburgh Academical Club: Edinburgh Academy Register 1824–1914; Class lists, 1844–6; Academical Club prizes, 1846; Syllabuses, 1840s; Minute Books of the Directors.
Edinburgh University Library: court and senate minutes; lists of graduates; letters of Fleeming and H.C. Jenkin.
Folkestone Library: Dunk undertaker's ledger f.1963/5/B3.
Glasgow University Library: Kelvin papers.
Harvard University, Houghton Library for Rare Books and Manuscripts: letter of Fleeming Jenkin, 7 March 1859.
Heriot-Watt University Archive: Minute Books of Directors.
Institution of Civil Engineers: Membership forms A and B.
Institution of Electrical Engineers: SC Mss 22, Sir William Preece papers; SC Mss 20, Sir Charles T. Bright papers; SC Mss 21, telegraphy press cuttings; SC Mss 51, patents re electric railways; membership application forms 1870–1901.
Institution of Mechanical Engineers: membership proposal forms.
Jenkin family mss, in the possession of Lieutenant Commander James Jenkin, RN, Binsted, Hampshire.
John Rylands University Library, Manchester: JA6/2/220–2, Jevons Papers.
National Library of Scotland: letters and photographs of Fleeming Jenkin; letters of Anne and H.C. Jenkin; Douglas papers; Acc. 9690, note of class excursion by R.L. Stevenson.
National Maritime Museum: Ms. 88/078, Hooper v. Elliot arbitration; TCM/6/5, Programme of Proceedings for laying the Franco-American cable; TCM/7/2, Register of submarine cables laid.
New York Public Library: Wheeler collection.
Parish of St Mary the Virgin, Stowting, Kent: burial register, 1813–present.
Public Record Office: Board of Trade records of companies (BT 31); service records of members of the Jenkin family (ADM 196); Chancery records, Thomson, Varley and Jenkin v. The Atlantic Telegraph Company (C16/533/T29); Treasury papers (T108).
Royal Society: Government Grant applications; biographical information on past fellows; printed List of Candidates, 1848–67.
Savile Club, London: records of members.
Science Museum Library: Muirhead collection; Henley papers.
Scottish Record Office: testament and inventory of Fleeming Jenkin, SC70/4/214; /1/244; /1/247.
University of St Andrews Library, Forbes mss.
Wirral Borough Archives: Birkenhead and Liverpool directories.
The Writers' Museum, Edinburgh: photograph of R.L. Stevenson.
Yale University Library, Beinecke Rare Book and Manuscript Library: Robert Louis Stevenson collection.

Contemporary Published Sources

W. Archer, 'Mr R.L. Stevenson's Memoir of Fleeming Jenkin', *Pall Mall Gazette*, 11 January 1888.

Electrical Trades Directory.

The Electrician.

Electricians' Directory.

W. Hawes, 'Report of the Committee on Technical Education', *Journal of the Society of Arts*, XVI, 24 July 1868, pp. 627–33.

W. Hole, *Quasi Cursores: portraits of the High Officers and Professors of the University of Edinburgh* (Edinburgh, 1884).

Minutes of Proceedings of the Institution of Civil Engineers.

R.S. Newall, *Observations on the Present Condition of Telegraphs in the Levant* (London: William Clowes & Sons, 1860).

P.G. Tait, 'Professor Fleeming Jenkin', *Nature*, 8 March 1888, pp. 433–5.

University College, London, *Annual Report* for 1868 and 1869.

J. Wagstaff Blundell, *The Manual of Submarine Telegraph Companies* (London: Rixon and Arnold, 1871).

Obituaries of Fleeming Jenkin

The Electrician, 19 June 1885, p. 97.

Minutes of Proceedings of the Institution of Civil Engineers LXXXII (1885), pp. 365–77.

Nature XXXII, 18 June 1885, pp. 153–4.

Journal of the Society of Telegraph Engineers XIV (1885), pp. 345–50.

Royal Society of Edinburgh Proceedings XIV (1886–88), pp. 117–19.

The Scotsman, 13 June 1885.

R.L. Stevenson, in *The Academy*, 20 June 1885

W. Thomson, 'Obituary Notices of Fellows Deceased', *Proceedings of the Royal Society*, XXXIX (1885), pp. i–iii.

Parliamentary Papers

Report of the Board of Trade Committee to inquire into the best Plan for Construction of Submarine Cables [Galton committee], 1860 [2744] LXII. 591.

Select Committee on Scientific Instruction [Samuelson Committee], 1867–68 [c. 432] XV. 1.

Royal Commission on Scientific Instruction and the Advancement of Science [Devonshire Commission], 1872 [c. 536] XXV. 92; 1874 [c. 958] XXII, 12–16; 1875 [c. 1279] XXVIII. 59; [c. 1297] [c. 1298] XXVIII. 331, 417; [c. 1363] XXVIII. 473.

Royal Commission on Technical Instruction, 1884 [c. 3981] XXIX. 493.

Educational Endowments (Scotland) Commission, 1884 [c. 3995] XXVII. 776.

Secondary Sources

R. Appleyard, *The History of the Institution of Electrical Engineers (1871–1931)* (London: IEE, 1939).

H. Barty-King, *Girdle Round the Earth: the story of Cable and Wireless* (London: Heinemann, 1979).

L.F. Bates, *Sir Alfred Ewing, a Pioneer in Physics and Engineering* (London: Longmans, Green and Co, 1946).

R.M. Birse, *Engineering at Edinburgh University: a short history, 1673–1983* (University of Edinburgh, 1983).

A.W. Bond, 'Edinburgh Exhibition, 1890: the work of Waller and Manville', *Modern Tramway*, January 1969, pp. 22–5.

B.A. Booth and E. Mehew (eds), *The Letters of Robert Louis Stevenson* (Yale University Press, 1994–95), 8 vols.

C. Bright, *Submarine Telegraphs: their History, Construction and Working* (New York: Arno, 1974 [1898]).

W.H. Brock, 'The Spectrum of Science Patronage', in G.L'E. Turner (ed.), *The Patronage of Science in the Nineteenth Century* (Leyden: Noordhoff, 1976).

A.D. Brownlie and M.F. Lloyd Prichard, 'Professor Fleeming Jenkin, 1833–1885: pioneer in engineering and political economy', *Oxford Economic Papers* XV (1963), pp. 204–16.

R.J. Cain, 'Telegraph Cables in the British Empire, 1850–1900', unpublished PhD thesis, Duke University, 1970.

J.A.V. Chapple and A. Pollard (eds), *The Letters of Mrs Gaskell* (Manchester University Press, 1966).

R.A. Chipman, 'Henry Charles Fleeming Jenkin' in C. C. Gillispie (ed.), *Dictionary of Scientific Biography* VII (New York: Charles Scribner's Sons, 1973), p. 93.

T.N. Clarke, A.D. Morrison-Low and A.D.C. Simpson, *Brass and Glass: Scientific Instrument Making Workshops in Scotland* (Edinburgh: National Museums of Scotland, 1989).

R.D. Collison Black (ed.), *Papers and Correspondence of William Stanley Jevons* (London: Macmillan, 1977).

R.D. Collison Black, 'Jenkin, Henry Charles Fleeming' in J. Eatwell, M. Milgate and P. Newman (eds), *The New Palgrave: a dictionary of economics*, II (London: Macmillan, 1987), pp. 1007–8.

S. Colvin, *Memories and Notes of Persons and Places, 1852–1912* (London: Edward Arnold & Co., 1921).

S. Colvin and J.A. Ewing (eds), *Papers Literary, Scientific, etc., by the late Fleeming Jenkin, F.R.S. , LL. D.* (London: Longmans, Green and Co., 1887), 2 volumes.

T. Constable, *Memoir of Lewis D.B. Gordon, FRSE, late Regius Professor of Civil Engineering and Mechanics in the University of Glasgow* (Edinburgh, 1877).

G. Cookson, 'Reconstructing a lost engineer: Fleeming Jenkin and problems of sources I – biographical sources' in A. Jarvis and K. Smith (eds), *Perceptions of Great Engineers II* (Liverpool: National Museums and Galleries on Merseyside, 1998).

P. Deane, *The Evolution of Economic Ideas* (Cambridge University Press, 1978).

A. Desmond and J. Moore, *Darwin* (London: Michael Joseph, 1991).

G.S. Emmerson, *Engineering Education: a Social History* (Newton Abbot: David and Charles, 1973).

A.W. Ewing, *The Man of Room 40: the life of Sir Alfred Ewing* (London: Hutchinson and Co., 1939).

J.A. Ewing, *An Engineer's Outlook* (London: Methuen, 1933).

H.M. Field, *The Story of the Atlantic Telegraph* (New York: Charles Scribner's Sons, 1893).

G. Gooday, 'Teaching telegraphy and electrotechnics in the physics laboratory: William Ayrton and the creation of an academic space for electrical engineering in Britain, 1873–1884', *History of Technology* XIII (1991), pp. 73–111.

K.R. Haigh, *Cable Ships and Submarine Cables* (1968).

R. Harré, *The Philosophies of Science* (Oxford University Press, 1985).

D.R. Headrick, *The Invisible Weapon: telecommunications and international politics, 1851–1945* (Oxford University Press, 1991).

C.A. Hempstead, 'An appraisal of Fleeming Jenkin (1833–85), electrical engineer', *History of Technology* XIII (1991), pp. 119–44.

C.A. Hempstead, 'Telpherage and problems of innovation', *IEE Proceedings* CXL(3) (May 1993), pp. 197–205.

C.A. Hempstead, 'Control systems and Fleeming Jenkin's telpherage', *Proceedings of the 21st annual meeting of the History of Technology Group, IEE* (London: IEE, 1994), pp. 52–71.

C.A. Hempstead, 'Reconstructing a lost engineer: Fleeming Jenkin and problems of sources II – technical sources: types, contents and interpretations' in A. Jarvis and K. Smith (eds), *Perceptions of Great Engineers II* (Liverpool: National Museums and Galleries on Merseyside, 1998).

C.A. Hempstead, 'The formation of electrical engineers in the United Kingdom, 1880–1900: constructing the foundations', *Proceedings of an International Symposium on Galileo Ferraris and the conversion of energy – development of electrical engineering over a century*, forthcoming.

Heriot-Watt University: from Mechanics' Institute to Technological University, 1821–1973 (Edinburgh: 1973).

W.E. Houghton (ed.), *The Wellesley Index of Victorian Periodicals, 1824–1900* (Toronto, 1966), I.

D.L. Hull, *Darwin and his Critics: the reception of Darwin's theory of evolution by the scientific community* (Cambridge, MA: Harvard University Press, 1973).

B.J. Hunt, 'The ohm is where the heart is: British telegraph engineers and the development of electrical standards', *Osiris* IX (1994), pp. 48–63.

B.J. Hunt, 'Scientists, engineers and Wildman Whitehouse: measurement and credibility in early cable telegraphy, *British Journal for the History of Science* XXIX (1996), pp. 155–69.

B.J. Hunt, 'Doing science in a global empire: cable telegraphy and electrical physics in Victorian Britain' in B. Lightman (ed.), *Victorian Science in Context* (University of Chicago Press, 1997).

B.J. Hunt, 'Insulation for Empire' in F.A.J.L. James (ed.), *Semaphores to Short Waves* (London: Royal Society of Arts, 1998).

T.W. Hutchison, 'The 'Marginal Revolution' and the decline and fall of English classical political economy' in R.D. Collison Black, A.W. Coats and C.D.W. Goodwin (eds), *The Marginal Revolution in Economics: interpretation and evaluation* (Durham, N.C. : Duke University Press, 1973).

J.V. Jeffery, 'The Varley family: engineers and artists', *Notes and Records of the Royal Society of London* LI (2), (1997), pp. 263–79.

D.J. Jeremy (ed), *Dictionary of Business Biography* (London: Butterworths, 1984–86), 5 vols.

C.G. Knott, *Life and Scientific Work of Peter Guthrie Tait* (Cambridge University Press, 1911).

G.L. Lawford and L.R. Nicholson (eds), *The Telcon story, 1850–1950* (London: Telegraph Construction Company, 1950).

E.V. Lucas, *The Colvins and their Friends* (London: Methuen, 1928).

C. Mair, *A Star for Seamen: the Stevenson family of engineers* (London: John Murray, 1978).

M. Magnusson, *The Clacken and the Slate; the story of Edinburgh Academy, 1824–1974* (London: Collins, 1974).

B. Matthews (ed.), *Papers on Acting* (New York: Hill and Wang, 1958).

F.J. McLynn, *Robert Louis Stevenson: a biography* (London: Hutchinson, 1993).

J.S. Mill, *Autobiography* (Oxford University Press, 1971).

J.B. Morrell, 'The Patronage of Mid-Victorian Science in the University of Edinburgh', in G.L'E. Turner (ed.), *The Patronage of Science in the Nineteenth Century* (Leyden: Noordhoff, 1976)

S.W. Morris, 'Fleeming Jenkin and *The Origin of Species*: a reassessment', *British Journal for the History of Science* XXVII (1994), pp. 313–43.

A.D. Morrison-Low and J.R.R. Christie (eds), *'Martyr of Science': Sir David Brewster, 1781–1868* (Edinburgh: Royal Scottish Museum, 1984).

R.S. Newall, *Facts and Observations relating to the invention of the submarine cable and to the laying of the first cable between Dover and Calais in 1851* (London: R.E. and F.N. Spon, 1882).

D. Newsome, *The Victorian World Picture: perceptions and introspections in an age of change* (London: John Murray, 1997).

W. Pole, *The Life of Sir William Siemens* (London: John Murray, 1888).

G.B. Prescott, *Electricity and the Electric Telegraph* (New York, 1877).

J. Rees, *The Life of Captain Robert Halpin* (Arklow: Dee-Jay, 1992).

W.H. Russell, *The Atlantic Telegraph* (Newton Abbot: David and Charles, 1972 [1865]).

M. Sanderson, *The Universities and British Industry, 1850–1970* (London: Routledge and Kegan Paul, 1972).

J.A. Schumpeter, *History of Economic Analysis* (Oxford University Press, 1954).

N.Q. Sloan, 'William Thomson's Inventions for the Submarine Telegraph Industry: a nineteenth-century technology program', unpublished Master of Liberal Arts thesis, Harvard University, 1996.

C. Smith and M.N. Wise, *Energy and Empire: a biographical study of Lord Kelvin* (Cambridge University Press, 1989).

W. Smith, *The Rise and Extension of Submarine Telegraphy* (New York: Arno, 1974 [1891]).

R.L. Stevenson, 'Memoir of Fleeming Jenkin', in Colvin and Ewing (eds) 1887, I, pp. xi–cliv.

R.L. Stevenson, 'Some College Memories', in *The Works of Robert Louis Stevenson* IX (1911).

R.L. Stevenson, 'Talk and Talkers', in *The Works of Robert Louis Stevenson* IX (1911).

J.A. Venn (ed.), *Alumni Cantabrigiensis, pt. II, 1752–1900*, (Cambridge University Press, 1947).

P. Vorzimmer, 'Charles Darwin and blending inheritance', *Isis* LIV (1963), pp. 371–90.

E. Whittaker, *A History of Economic Ideas* (New York: Longmans, Green and Co., 1947).

D.B. Wilson (ed), *Catalogue of the Manuscript Collections of Sir George Gabriel Stokes and Sir William Thomson in Cambridge University Library* (Cambridge University Press, 1976).

D.B. Wilson (ed.), *The Correspondence between Sir George Gabriel Stokes and Sir William Thomson, Baron Kelvin of Largs* (Cambridge University Press, 1990), 2 volumes.

R.B. Wilson (ed), *Sir Daniel Gooch: memoirs and diary* (Newton Abbot: David and Charles, 1972).

Y.Y., *Robert Louis Stevenson: A Bookman Extra Number,* London, Hodder and Stoughton, 1913).

Index